Bryan Higgins

Experiments and Observations Made with the View of Improving the Art of Composing and Applying Calcareous Cements and of Preparing Quick-Lime

Bryan Higgins

Experiments and Observations Made with the View of Improving the Art of Composing and Applying Calcareous Cements and of Preparing Quick-Lime

ISBN/EAN: 9783337813536

Printed in Europe, USA, Canada, Australia, Japan

Cover: Foto ©Thomas Meinert / pixelio.de

More available books at **www.hansebooks.com**

EXPERIMENTS

AND

OBSERVATIONS

MADE

With the VIEW of IMPROVING the ART of COMPOSING and APPLYING

CALCAREOUS CEMENTS

AND

Of preparing Quick-lime:

THEORY of thefe ARTS;

AND

Specification of the AUTHOR's cheap and durable CEMENT, for Building, Incruftation or Stuccoing, and artificial Stone.

By *BRY. HIGGINS*, M. D.

L O N D O N:

Printed for T. CADELL, oppofite *Catherine-Street*, in the *Strand*.

M,DCC,LXXX.

CONTENTS.

SECTION I.
ORIGIN of these experimental Enquiries, — Page 1

SECTION II.
Experiments and Observations on Lime-stone and Lime, — — 3

SECTION III.
Remarks on the phlogisticated Air which appeared in some of the foregoing Experiments, — — 14

SECTION IV.

Experiments shewing that Lime is better for Mortar as it retains less acidulous Gas, and shewing some of the Causes of the Imperfection of common Mortar, — 17

SECTION V.

Experiments shewing how quickly Lime imbibes acidulous Gas, and is injured by Exposure to Air: Practical Inductions, &c. 29

SECTION VI.

Experiments and Observations made to determine whether Mortar be the better for being long kept before it is used, — 37

SECTION VII.

Of the Depravation of Mortar by the common Method of using the Water; and of the Use of Lime Water, — 43

CONTENTS.

SECTION VIII.

Experiments made with a View to approximate the best Proportions of Lime Sand and Water, for Mortar, — 46

SECTION IX.

Theory of the Induration dependent on the Proportions of Lime and Sand in Mortar, and Observations on the bad Effects of the vulgar Proportions of these, — 54

SECTION X.

Experiments on old Cements, authorizing the Proportion lately recommended of Lime and Sand, — — 60

SECTION XI.

Experiments and Observations shewing the Agency of acidulous Gas in the Induration of Mortar, and Circumstances which impede or promote it. Practical Inferences, 62

SECTION XII.

Experiments shewing the best Kinds and Mixtures of Sand, and the best Method of using the Lime Water, in making Mortar,   78

SECTION XIII.

Experiments shewing the Effects of finest Sand and quartose Powder in Mortar: Observations on the finest calcareous Cements. Practical Precepts, — 97

SECTION XIV.

Experiments made on a larger Scale with our best Mixture of Sands Lime Water and Lime, — — 112

SECTION XV.

Experiments shewing the integrant Parts of Gravel, the Choice and Preparation of it; and the Effects of Clay, Fuller's Earth, and Terras, in Mortar, — 119

CONTENTS.

SECTION XVI.

Experiments shewing the Effects of Plaister Powder, Alum, Vitriolic Acid, of some metallic and earthy Salts, and of Alkalies, in Mortar. Practical Inferences, — 126

SECTION XVII.

Experiments shewing the Effects of skimmed Milk, Serum of Ox-Blood, Decoction of Lintseed, Mucilage of Lintseed, Olive Oil, Lintseed Oil, and Resin, in Mortar; and the Effect of painting calcareous Incrustations, — — 133

SECTION XVIII.

Experiments shewing the Effect of Sulphur, introduced by different Methods, in Mortar, 139

SECTION XIX.

Experiments shewing the Effects of Crude Antimony, Regulus of Antimony, Lead Matt, Potter's Ore, White Lead, Arsenic, Orpiment, Martial Pyrites and flaked Mundic, in Mortar, — 143

SECTION XX.

Experiments shewing the Effects of Iron Scales, washed Colcothar, native Red Ochres, Yellow Ochres, Umber, Powder of coloured Fluor, coloured Mica, Smalt, and other coloured Bodies, in Mortar. Advices concerning coloured Incrustations, Inside Stucco, and damp Walls,   148

SECTION XXI.

Experiments shewing the Effects of common Wood-ashes, calcined or purer Wood-ashes, elixated Ashes, Charcoal Powder, Sea Coal-ashes, and powdered Coak, in Mortar; and Observations on their integrant Parts, and the Differences between them and the Powders of other Bodies, —— 160

SECTION XXII.

Experiments shewing the Effects of white and grey Bone-ashes, and the Powder of charred Bones; and Theory of the Agency of these in the best calcareous Cements, 172

SECTION XXIII.

The Specification made in Consequence of Letters Patent, illustrated with Notes,   Page 184

SECTION XXIV.

Experimental Comparisons of Chalk Lime with Stone Lime. Advices to the Manufacturers of Chalk Lime, concerning the Art of rendering it equal, if not superior, to Stone Lime, for the Purposes of Builders Soap-Boilers and Sugar-Bakers,   206

SECTION XXV.

Directions to the Houses already stuccoed with the new Cement. Observations on the Objections of certain Artists; on the Cementitious Works of the Romans; on the experienced and unequalled Duration of such Cements: on the Cements of Loriot and others; and on certain Uses of the Author's Cement,   214

SECTION I.

AMONGST the inftances and experiments produced in my public Courfes of Chemiftry in 1774, to illuftrate my notions of the polarity of matter, divers mixtures of lime fand and water, were particularly confidered: and thefe being preferved and methodically arranged, according to the plan of this fchool, foon fuggefted to me an enquiry which I have profecuted with great attention ever fince that time.

As the ftrength and duration of our moft ufeful and expenfive buildings depend chiefly on the goodnefs of the cement with which they are conftructed, I looked to the improvement of mortar as a fubject of great importance, in this country particularly, where the weather is fo variable and trying, and the mortar commonly ufed is fo bad, that the timbers of houses last longer than the walls, unless the mouldering cement be frequently replaced by pointing. But seeing that many years are requisite for the greatest degree of induration which cementitious mixtures like mortar can acquire, or for our discovering the imperfections of them; and that the life of man is too short to allow any considerable improvements of them to be derived from such experiments as had hitherto been made, I resolved in the beginning of the year 1775 to investigate more closely than I had hitherto done, the principles on which the induration and strength of calcarious cements depend; not doubting that this would lead me by an untried path to recover or to excel the Roman cement, which in aqueducts and the most exposed structures has withstood every trial of fifteen hundred or two thousand years.

SECTION II.

Experiments and Observations on Lime-stone and Lime.

I Had already learned from the chaste and philosophic productions of Dr. Black, that calcarious stones which burn to lime, contain a considerable quantity of the elastic fluid called fixable air or acidulous gas, which in combination with the earthy matter forms a great part of the mass and weight of these stones; and that the difference between lime-stone or chalk and lime, consists chiefly in the retention or expulsion of this matter.

Expecting to learn something further relative to lime, and particularly, to discover the cause of the differences which appear in cements made with different kinds of lime, I made the following experiments.

I procured specimens of different kinds of lime-stone and chalk, and breaking them into

into small fragments, I burned them in crucibles lined with lime to prevent the pieces from touching their crucibles and vitrifying at those surfaces which lay next to them: I likewise burned the like specimens in crucibles perforated to admit a free current of air through them; and lastly, I exposed three pounds of either specimen to a graduated fire in an earthen retort which was barely sufficient to hold this quantity, and whose neck I lengthened by fitting to it a glass conical tube luted at the juncture with four parts of lime one of fine sand and as much dissolved glue such as the carpenters use, as was sufficient to form a paste; having found this luting to hold fast and to be impervious to any elastic fluid or liquor expelled in such processes. I immersed the extremity of the glass tube in mercury, and inverting a bottle filled with mercury over the extremity of the tube, I received whatever water or elastic fluid was expelled from the calcarious stones by the fire, and I measured the quantity of these by instantly applying a fresh bottle as soon as the former was filled. When all the water was expelled, or when I knew the quantity of it contained in the

lime-

lime-ftone or chalk, I ufed a bafon of water and bottles filled with water, making allowance for the matter imbibed by the water.

To avoid a tedious detail of particulars which do not immediately relate to the chief object of this effay, I fhall only mention fummarily the moft pertinent obfervations which thefe and other experiments afforded; endeavouring that the terms in which I fhall deliver thefe obfervations fhall defcribe the experiments fufficiently for thofe who are acquainted with modern chemiftry.

OBSERVATION I.

LIME-STONE or chalk heated only to rednefs, in a covered crucible, or in a perforated crucible through which the air circulates freely, loofes only about one-fourth of it's weight, however long this heat be continued. The fort of lime fo formed effervefces confiderably in acids, flakes flowly and partially to a powder which is not white, but is grey or brown, and heats but little in flaking.

IN defcribing heats I do not regard the heat in particular parts of the fuel, but only

only that which the bodies themselves are made to conceive equally through their whole mass, whether they be in vessels which defend them considerably from the action of the fire, or fully exposed to it by their immediate contact with the fuel.

Observation. 2.

Lime-stone or chalk exposed to a heat barely sufficient to melt copper, whether in a perforated crucible or otherwise, loses about one third of its weight in twelve hours, and very little more in any longer time. This lime effervesces but slightly in acids; it heats much sooner and more strongly than the foregoing, when water is sprinkled on it, and it flakes more equably and to a whiter powder. In a variety of trials, this lime appeared to be in the same state with the best pieces of lime, prepared in the common lime-kilns. For the quantities of acidulous gas obtainable from both by a stronger heat, or in solution, were nearly equal; they flaked in equal times, with the same phenomena, and to the same colour and condition of the powder.

OBSERVATION. 3.

The lime burned in perforated crucibles, or in the naked fire, is whiter than that burned in common crucibles covered, in which cafe the air has not fo free accefs to it; altho' the lofs of weight be the fame in both; but this latter kind of lime, in flaking, affords as white a powder as any other which has loft equally of its weight. Whatever portion of phlogifton it retains to produce this dufky colour, is either detached in the flaking, or does not fenfibly affect the lime in any ufe, to which I applied it.

OBSERVATION. 4.

When dry chalk or lime-ftone is ufed, in the procefs above defcribed for making lime in clofe veffels, and for examining the matter which is expelled by fire, the quantity of water obtainable from it by any heat, is fo inconfiderable as to deferve no notice in our menfuration of that matter.

OBSERVATION. 5.

Chalk or lime-ftone heated gradually in thefe clofe veffels, lofes very little acidulous gas until it begins to redden: after this the elaftic fluid iffues from it the quicker as the heat is made greater, and continues to iffue until the retort glows with a vivid white heat fufficient to melt fteel.

OBSERVATION. 6

FORTY-EIGHT ounces of chalk yield twenty-one ounces of elaftic fluid, the firft portions of which are turbid as they iffue, but foon become clear without lofs of bulk, by the condenfation of the watery vapour: the remaining portions iffue tranfparent and invifible. One thirty-fixth of this elaftic fluid, and fometimes much more of it, is phlogiftic air, the remainder is pure acidulous gas.

OBSERVATION. 7

The refiduary lime of forty-eight ounces of chalk, urged with fuch heat to the total expulfion of the elaftic fluids, weighs only twenty-feven ounces, whilft it is red hot. When it cools it weighs more by reafon of the air which it imbibes as the fire efcapes from it.

OBSERVATION 8.

When no more heat is employed than is neceffary for the expulfion of thefe elaftic fluids, the refiduary matter is found contracted fenfibly in volume, and is good lime, tho' not fo white as lime prepared in the ufual way. With water it flakes inftantly, grows hiffing hot and perfectly white. The flaked powder is exceedingly fine, except in thofe parts of the lime which lay in contact with the retort, which are always fuperficially vitrified, becaufe clay and lime promote the vitrification of each other.

OBSERVATION 9.

The lumps of this lime, immerfed in lime-water, or boiling water, to expel the air which fuch fpongy bodies imbibe in cooling, diffolve in marine acid without fhewing any fign of effervefcence.

OBSERVATION 10.

Lime-stone or chalk gradually heated in a crucible, or on the bed of a reverberatory furnace, or in contact with the fuel in a wind furnace, does not become perfectly noneffervefcent and fimilar to the lime laft defcribed scribed, in flaking inftantly, and growing hiffing hot when water is fprinkled on it, until it has, after a ftrong red heat of fix or eight hours, fuftained a white heat for an hour or more. I underftand by a white heat, that which is fufficient to melt caft iron compleatly.

OBSERVATION. 11.

LIME-STONES heated fufficiently to reduce them to lime which flakes inftantly with the figns above defcribed, and which is perfectly noneffervefcent, do not in general lofe fo much of their weight as chalk-ftone does, under the like treatment. Some lime-ftones lofe little more than a third of their weight. Thofe which lofe the moft, flake the quickeft and to the fineft powder; and thofe which lofe the leaft, flake the floweft and to a gritty powder compofed of true lime and particles chiefly gypfeous.

OBSERVATION. 12.

THE quantity of gypfum, or of other earthy matter in well burned lime, is difcoverable by weak marine acid; for this diffolves and

and washes away the lime, leaving the gypsum to be measured when dry, the part of
the gypsum which dissolves being too small
to deserve any attention; and if any other
earthy matter or any saline matter existed in
the lime-stone, it vitrifies with part of the
calcarious matter in the heat necessary for
making noneffervescent lime, and is separable
by the means last mentioned, and even by a
fine sieve in most instances.

Observation. 13.

When lime-stone or chalk is suddenly
heated to the highest degree above described,
or a little more, it vitrifies in the parts which
touch the fire vessels, or furnace, or fuel, and
the whole of it becomes incapable of slaking
freely or acting like lime. Lime-stone is
the more apt to vitrify in such circumstances, as it contains more gypseous or argillaceous particles; and oyster-shells or cockle-
shells vitrify more easily than lime-stone or
chalk, when they are suddenly heated;
which I impute to their saline matter; for
when they are long weathered, they do not
vitrify so easily.

OBSERVATION. 14.

THE agency of air is no further neceſſary in the preparation of lime, than as it operates in the combuſtion of the fuel.

OBSERVATION. 15.

CALCARIOUS ſtones acquire the properties of lime in the moſt eminent degree, when they are ſlowly heated in ſmall fragments until they appear to glow with a white heat, when this is continued until they become noneffervefcent, but is not augmented. The art of preparing good lime conſiſts chiefly in theſe particulars.

OBSERVATION 16.

THAT lime is to be accounted the pureſt and fitteſt for experiment, whether it be the beſt for mortar or not, which flakes the quickeſt and heats the moſt in flaking, which is whiteſt and fineſt when flaked, which when wetted with lime-water diſſolves in marine acid or diſtilled vinegar without effervefcence, and leaves the ſmalleſt quantity of reſiduary undiſſolved matter.

OBSER-

OBSERVATION 17.

The quick flaking, the colour of the flaked powder, and the former acid, are the moſt convenient, and perhaps the beſt teſts of the purity of lime. The whiteneſs denotes the lime to be free from metallic impregnation; the others ſhew any imperfections in the proceſs of burning, and the heterogeneous matter inſeparable from the calcareous earth by burning.

SECTION

SECTION III.

Remarks on the Phlogisticated Air which appeared in some of the foregoing Experiments.

AS phlogisticated air had not been noticed in any experiment heretofore made on chalk or lime-stone, I resolved to examine the elastic fluid detached from them in the usual method.

I extricated several gallons of elastic fluid from chalk, during the solution of it in marine acid diluted largely with water; and after agitating this elastic fluid with the necessary quantity of water, and sometimes with lime-water, until all the acidulous gas was imbibed by them, I found a residue consisting of common air, which was about one twenty-eighth of the bulk of the acidulous gas, in some trials, in others it was much less.

As I have not had time to examine lime-stones in the same manner, or to prosecute this

this subject by other experiments, and as it does not appear sufficiently interesting in our present enquiry concerning calcareous cements, I must content myself with offering a conjecture concerning it.

THE air which is extricated during the solution of chalk, seems to be that which chalk, like other porous bodies, imbibes by capillary attraction; and it retains its proper character, because all the phlogistic matter of chalk is held in the solution. It may happen likewise that some air escapes from the water whilst it imbibes the acidulous gas, which it attracts more forcibly; and this air from the water may contribute to the bulk of that which appears in the solution of calcarious bodies. But whilst chalk is deprived of its acidulous gas by the action of fire, the air which was held in its pores, and which attracts phlogiston, is expelled in combination with phlogiston, and consequently in the form of phlogistic air; and the air contained in the pores and in the cavity of the retort, contributes to the bulk of the phlogistic air obtainable in this manner.

THIS

This conjecture appears the more probable when we confider, that the quantity of air imbibed by porous bodies is much greater than it appears in any experiments made with the air pump; as I fhewed in my public courfe of chemiftry in 1776, by the great increafe of weight, which red hot charcoal acquired in cooling in veffels into which nothing ponderable but air was admitted. The fame attractive powers which draw air into bodies, and condenfe it in them, refift the expanfion and efcape of it in the void; and detain, in fuch a fituation of the bodies, that quantity, whofe repulfive powers are counterpoifed by the attractive powers.

SECTION

SECTION IV.

Experiments shewing that Lime is better for Mortar as it retains less acidulous Gas, and shewing some of the Causes of the Imperfection of common Mortar.

ON divers confiderations it appeared to me, that the perfection of lime for mortar confifts chiefly in the total expulfion of the acidulous gas; but to be better fatisfied of the truth of this opinion, I made feveral parcels of mortar, the defcription of which will be abridged by obferving in this place concerning all of them, that the fand employed was coarfe Thames fand, fuch as I ufe in my fand baths; that the lime was flaked as foon as it cooled after being burned, and with the fmalleft quantity of water neceffary for this purpofe; that it was fifted through a fine brafs wired fieve as foon as it was fully flaked; and that each parcel of mortar was beaten and brifkly formed with the quantity of water which was barely fufficient to give it the ufual confiftence, which

C quantity

quantity I shall express in the pharmaceutical method by q. s.

SPECIMEN 1.

Sand — —	3
Purest stone lime, described in Obs. 10, and 16, sect. 2, —	1
Water, q. s.	

2.

Sand, — —	6
Purest stone lime, last mentioned,	1
Water, q. s.	

3.

Sand, — —	3
Purest chalk lime, described in obs. 10 and 16, sect. 2. —	1
Water, q. s.	

4.

Sand, — —	6
Purest chalk lime, last mentioned,	1
Water, q. s.	

5. Sand

5.

Sand — — 3
Stone lime of the beft kind, except that it was burned no further than is expreffed in the fecond obfervation of the fecond fection, — 1
Water, q. f.

6.

Sand, — — 6
The laft mentioned ftone lime, 1
Water, q. f.

7.

Sand, — — 3
Chalk lime defcribed in the fecond obfervation of the fecond fection, 1
Water, q. f.

8.

Sand, — — 6
The laft mentioned chalk lime, 1
Water, q. f.

9.

Sand, — — 3
Imperfect lime, defcribed in obf. 1, fect. 2, but formed of the beft lime-ftone, 1
Water, q. f.

10. Sand,

10.

Sand, — — 6
The foregoing lime, — 1
Water, q. f.

11.

Sand, — — 3
Imperfect chalk lime described in obf. 1,
   sect. 2, — — 1
Water, q. f.

12.

Sand, — — 6
The last mentioned lime — 1
Water, q. f.

THE lime of the 9, 10, 11, and 12 specimens was flaked whilst it was hissing hot, in a covered vessel; because it would not flake sufficiently when it was suffered to cool before the water was sprinkled on it, or when it's heat was soon dissipated by a free exposure to the air and hasty evaporation of the water: And as this lime required several hours to flake, I put it into a bottle as soon as it was cool, and kept it well stopped for twenty-four hours before I sifted it.

At

At the same time I made two specimens of mortar with common chalk lime and sand in the foregoing manner.

Each specimen was spread as soon as it was made, to the thickness of half an inch on a plain tile previously soaked in water; the tiles were numbered and kept close by each other in an airy part of my elaboratory until the mortar was dry, and then they were equally exposed, standing upright, in a place where the air, sun and rain had free access to them.

In the course of fourteen or fifteen months these specimens afforded me a great deal of information, which will be noticed in due time; even in the first six months they shewed me clearly that lime is the better for mortar as it is more perfectly freed from acidulous gas. For when the comparison was made between specimens of mortar consisting of the same quantities of lime and sand, I found that the mortar made with well burned non-effervescent lime, hardened sooner and to a much greater degree, than mortar made with common lime or my stone or chalk lime burned

burned in the manner expreſſed in the ſecond obſervation of the ſecond ſection: and the ſpecimens made with the ſtone or chalk lime which was leaſt burned, were incomparably worſe than any of the others; for they never acquired any conſiderable hardneſs, and they mouldered in the winter, the ſooner as they contained more of the lime and cracked more in drying.

I obſerved that the ſpecimens which contained the ſmaller quantities of well burned lime cracked much leſs than the others, or not at all; that they adhered to the tiles more firmly, and were leſs injured by freezing; but as the ſpecimens made with an exceſs of the beſt burned lime were not more cracked than thoſe made with equal quantities of the other kinds of lime, and as I could diſtinguiſh the imperfections ariſing from the exceſs of lime, from thoſe which proceeded from the bad quality of it, I was ſatisfied that the lime which is moſt compleatly burned is the beſt for mortar.

CONSIDERING the heat, which I found neceſſary to extricate the laſt portions of acidulous gas from chalk or lime-ſtone, to be much greater

greater than what is ever excited in making lime in this country or elsewhere, so far as I had observed or could learn from others; I suspected that the lime commonly used in building is seldom or never sufficiently burned.

On repeated trials of several specimens of such lime, I found this suspicion to be well founded, for they all effervesced and yielded acidulous gas, more or less, during the solution of them, and slaked slowly in comparison with well burned lime.

To render the effervescence conspicuous, a strong acid ought to be used, because the quantity of water in a diluted acid, retains a proportional share of the acidulous gas, and a certain quantity will retain the whole of it, and prevent the effervescence; because the effervescence depends on the escape of the elastic fluid out of the solution. This is exemplified in the mixture of diluted vitriolic acid with the diluted solution of salt of tartar, lately recommended as a medicine by Dr. Hulme: for these solutions mix without effervescence; although a more concentrated solution of the alkali mixed with vitriolic acid effervesces violently.

By several experiments, the relation of which is not necessary for those who are instructed in chemistry, and would be uninteresting to others, I found that the chalk lime used in London, when taken as fresh as it can be had at the lime-wharf near Blackfriars-bridge, consists of pieces which being the best burned contain, especially in their central parts, about one-twentieth of their weight of acidulous gas; of others which contain more; and of others which retain near half their original quantity; that these last are easily discoverable by their specific gravity and hardness; and that this is the part of our common lime, which flakes the latest and of the duskiest colour, or which never flakes at all.

On a peck of this lime I sprinkled water, endeavouring to flake it equably by throwing the most water on those pieces which required it most. After the lime had stood a quarter of an hour, to flake, I sifted it through a sieve whose apertures were squares of one sixteenth

sixteenth of an Inch; and then measuring the part which could not pass through the sieve, I found it to be about one fifth part of a peck.

I sprinkled boiling water on this coarse part, and put it in a close vessel in a warm place to accelerate the flaking of it.

I made a parcel of mortar with one part of the sifted lime and three of sand with a sufficient quantity of water; and another parcel with one part of the lime six of sand and the necessary quantity of water; and I tried them upon tiles in the manner already related, in the month of April, the weather being dry.

The foregoing coarse portion of the lime, after three hours, was flaked in several parts to a greyish powder, and I could perceive that more of it would flake in a longer time. I anticipated this by reducing the unflaked part to powder and mixing them together,

WITH

WITH this powder and fand and water in the foregoing proportions I made two fpecimens of mortar, and expofed them as I had done by the former.

IN a few months it appeared that the fpecimens laft mentioned fcarcely deferved the name of mortar; whilft thofe made with the firft flaked part of the lime, were but little inferior to the beft fpecimens made with the fame proportions of chalk lime and fand.

THESE experiments confirmed me in the opinion that lime is the better for mortar as it is freer from acidulous gas; they fhewed one of the caufes of the badnefs of our common mortar; and how to manage ill-burnt lime, when better cannot be had.

THE workmen ufually flake the lime mixed with the fand or gravel in great heaps, and do not fkreen it until the moft ufeful part is debafed by that which flakes after five or fix hours or more, and which is little better than fo much powder of chalk. But if they would fkreen the lime in about half an hour

hour after the water is thrown on it, the mortar would be much better, although the quantity of lime in it fhould be much lefs; for I obferved in all the foregoing fpecimens, that thofe which contained the fmalleft quantity of lime were the beft; and this quantity is much fmaller than is ufually employed in making mortar.

These remarks are applicable to mortar made with ftone-lime; though the ftone-lime be generally better than the chalk-lime ufed in London, becaufe they are obliged to burn it better, as it will not flake otherwife.

In the brief relation of thefe experiments I have taken no notice of the flint kernels which frequently occur in chalk-lime, or of the other ftoney maffes different from the calcarious, which are found in lime-ftone; becaufe I took care that they fhould not lead me into any errors.

When firft I noticed the quantity of chalk-lime which flakes lateft or not at all, I fufpected that this difference might in fome degree be owing to the admixture of argillaceous

or

or other matter; but on trying thefe parts in acids, and after burning feveral fpecimens of them, I was convinced that the only impediment to their flacking confifted in their not being fufficiently burnt in the kiln.

SECTION V.

Experiments shewing how quickly lime imbibes acidulous gas, and is injured by Exposure to Air: Practical Inductions, &c.

IT was already known that lime exposed to air gradually loses those characters which chiefly distinguish lime from whiting or powder of chalk, and that it resumes the acidulous gas which had been expelled from it in burning. But as I was desirous to know in what measure or time these changes take place, and in what circumstances they are accelerated or retarded, I made the following Experiments.

On the 22d of August 1776 I exposed two pounds avoirdupoise of well-burned noneffervescent chalk lime, in fragments of the size of a walnut spread on a board, in a dry unfrequented room. I exposed the same quantity of this lime, at the same time and in the same manner, in a passage through which there was a constant current of air; and I

put

put the fame quantity of this lime, in fragments of the fame fize, in a box which might hold as much more of it, and placed the box loofely covered with its lid, clofe by the firft portion of lime.

In 24 hours the fuperficial lumps of the firft parcel cracked in fome parts a little, thofe of the fecond cracked more, thofe of the third were not vifibly altered. In forty-eight hours the firft parcel cracked fo much as to fall into fmaller fragments on being moved; and thefe were reducible to powder by preffing them between the fingers: The fecond parcel underwent the like or rather a greater change, for it was more cracked and friable: The third now begun to crack in the fuperficial parts.

On weighing them, I found that the firft parcel weighed two pounds five ounces, the fecond two pounds fix ounces and one drachm, the third two pounds one ounce ten drachms: I then returned them to their former ftations.

In six days the first parcel weighed two pounds ten ounces seven drachms; the second two pounds twelve ounces one drachm; the third two pounds four ounces eight drachms.

In twenty-one days the first parcel weighed three pounds one drachm; the second three pounds two ounces one drachm and a half; the third two pounds six ounces eight drachms.

During this increase of weight the fragments split into smaller pieces, but did not fall into powder, except in a small part of them, or when they were handled.

By similar experiments made on well burned stone lime I found that this imbibes matter from the air nearly in the same manner as chalk lime, but rather more slowly; which I think is owing to its closer texture.

On exposing common chalk or stone lime in the same way, I find that it increases in weight much less and more slowly.

To discover the quantity of water which the lime imbibed from the air, and which contributed to this increase of weight, I put each parcel in a glass retort; and adjusting to it my apparatus whereby all that is condensible is saved, whilst elastic fluids are at liberty to escape, I found that the quantity of water contained in each parcel of lime, was nearly in some, and in others accurately one-twenty-fourth of the gained weight, the remainder of the weight gained was of acidulous gas mixed with a little air, which latter I do not reckon, because it was already weighed in the lime.

If a glass bottle be filled with fragments of well-burned chalk lime, or stone lime, or shell lime, and well stopped with a ground glass stopple slightly waxed where it fits the neck of the bottle, the lime will remain unaltered in weight, or in any other known particular, for a year or two; as I have repeatedly experienced: even the phosphorescence of lime is thus preserved in its full lustre, for a year or more.

Thus it appeared that well burned lime imbibes

imbibes acidulous gas from the air, the sooner as it is the more fully expofed to it: that lime imbibes this matter from the open air, the more greedily as it is more perfectly deprived of it previous to the expofure: that lime cannot be long preferved unaltered in any veffels which are not perfectly air-tight, but may be kept uninjured for any time in air-tight veffels filled with it: that chalk lime, by reafon of its fponginefs, or by fome other condition of it, requires to be kept lefs expofed than ftone lime, and well burnt lime lefs expofed than common lime, to render the depravation of them equal in equal times: that if acidulous gas imbibed by lime previous to its being ufed in mortar, be as injurious to the mortar, as the acidulous gas retained in equal quantity by ill-burned lime is, lime grows the more unfit for mortar every hour that it is kept expofed to air, whether in a heap, or in cafks pervious to air.

I THINK moreover that thefe experiments fhew that lime undergoes thefe changes by expofure, much quicker than has been fufpected; fince well burned chalk lime kept in a dry room, imbibes near a pound of acidulous gas in three weeks, in the summer season.

Not to trust to theory what I could prove by experiment, I did not rest satisfied with the observations and reasons which might persuade one that lime, which has imbibed some acidulous gas, is as unfit for the uses now under consideration, as lime which retains an equal quantity of the like matter by reason of the deficiency of heat in burning it.

I tried parcels of well burned chalk and stone lime, some of which were used fresh, others exposed two days, others six days, others twenty-one days, in the same circumstances; by making several specimens of mortar with them, and exposing the specimens in the manner already related: and in a few months I was satisfied that the specimens made with fresh lime were the hardest and best, and that the others were worse as the lime of them had been longer exposed: for those made with the lime which had been exposed three weeks and had gained four or five ounces to each pound, were so

easily

eafily cut or broke, fo much affected by moifture and drying, and fo liable to break off from the tiles, as to be utterly unfit for the ordinary ufes of mortar.

AFTER this there remained no doubt that lime grows worfe for mortar every day that it is kept in the ufual manner in heaps or in crazy cafks; that the workmen are miftaken in thinking that it is fufficient to keep it dry; that lime may be greatly debafed without flaking fenfibly; and that the fuperficial parts, of any parcel of lime, which fall into fmall fragments or powder without being wetted, and merely by expofure to air, are quite unfit for mortar; fince this does not happen until they have imbibed a great deal of acidulous gas.

I NOW faw more clearly another caufe of the imperfection of our common cements. The lime being expofed a confiderable time before it is made into mortar, and drinking in acidulous gas all the while, the quicker as it is the better burned, is incapable of acting like good lime, when it is made into mortar; and often approaches to the condition of whiting, which with fand and water makes a friable perishable mass, however carefully it be dried. In London particularly they use lime which is burned, at the distance of ten or twenty miles or more, in Kent and elsewhere, with an insufficient quantity of fuel. This lime remains in the kiln, to which the air has access, for many hours after it is burned. It is exposed for some days in the transportation, and on the lime-wharfs; and it undergoes further exposure and carriage before the artist flakes it for mortar. It is no wonder that the London mortar is bad, if the imperfection of it depended solely on the badness of the lime; since the lime employed in it, is not only bad when it comes fresh from the kiln, but becomes worse before it is used, and when flaked is as widely different from good lime, as it is from powdered chalk.

SECTION

SECTION VI.

Experiments and Observations made to determine whether Mortar be the better for being long kept before it is used.

I AM generally disposed to think that there is some good reason for any practice which is common to all men of the same trade, although it may not be easily reconcilable to the notions of others: and seeing that the builders slake a great quantity of lime at once, more than they can use for some days, and that all those, whom I conversed with, esteemed mortar to be the better for being long made before it is used; and that plaisterers particularly follow this opinion in making their finer mortar or stucco for plaistering within-doors; I was desirous to discover the grounds of these measures so repugnant to the notions gathered from the foregoing experiments and others.

In the month of March 1777 I made about a peck of mortar, with one part of the

freshest

freſheſt and beſt chalk lime flaked, ſix parts of ſand, and water q. ſ; for in a great number of experiments, I obſerved that this proportion of lime was better than any larger which I had tried, or which the workmen obſerve in making mortar.

I FORMED the mortar into an hemiſpherical heap on the paved floor of a damp cellar, where it remained untouched twenty-four days. At the expiration of this time, I found it hardened at the ſurface; but moiſt, and rather friable or ſhort than plaſtic in the interior parts of it.

I BEAT the whole of it with a little water to its former conſiſtence; and with this mortar and clean new bricks, I built a wall eighteen inches ſquare and half a brick in thickneſs, in a workman-like manner. On the ſame day I made mortar of the ſame kind and quantities of freſh chalk lime and ſand, tempered in the ſame manner; and I built a wall with it, like the former, near it, and expoſed equally to the weather.

I EXAMINED the mortar in the joints of thefe walls every fortnight, by picking it with a pointed knife, and could perceive a very confiderable difference in the hardnefs of them; the mortar which was ufed frefh being invariably the hardeft.

AT the expiration of twelve months, in pulling thefe walls to pieces, and by feveral trials of the force neceffary to break the cement and feparate the bricks, I found the mortar which had been ufed quite frefh, to be harder and to refift fracture and the feparation of it from the bricks, in a much greater degree than the other fpecimen.

CONSIDERING that mortar expofed in the foregoing manner, muft imbibe fome acidulous gas, though not fo much, perhaps, as the dry and fpongy lumps of lime drink in, during the fame time; that the additional quantity of water neceffary in beating it up the fecond time, muft have introduced more of the like matter, as all native waters contain fome quantity of it; that the frefh expofure in the laft mentioned agitation of the mortar muft have contributed fomething to the fame effect;

effect; and lastly that the event of this experiment coincided with the notions already derived from others; I concluded that mortar grows worse every hour that it is kept before it is used in building, and that we may reckon as another cause of the badness of common mortar, that the workmen make too much at once, and falsly imagine that it is not the worse but better for being kept some time.

Having in consequence of these observations had a great deal of conversation with workmen on this subject, I could perceive the origin of this error.

Some portions of every kind of lime used in this country, do not flake freely, by reason of their not being sufficiently burned, or of the admixture of gypseous or argillaceous matter; and these, like marle, flake in time, though not so quickly as the purer lime.

The plaisterers, who use a finer kind of mortar made of sand and lime, observe that their plaister or stucco blisters, when it contains

tains fmall bits of unflaked lime; and as their purpofe is to work their ftucco to a fmooth furface, and to fecure it from cracking, or any fuch roughnefs as would be occafioned by the flaking or mouldering of bits of calcareous matter in the face of it; and as the hardnefs of the ftucco is not their chief object; they very properly keep their mortar a confiderable time before they ufe it, to the end that the bits of imperfect lime, which paffed through the fkreen, may have time to flake thoroughly.

It appears to me that there is another reafon, which the workmen do not notice, for their procefs. Lime foon imbibes fo much acidulous gas from the air, as to be increafed in bulk, and in weight beyond the half of its former quantity; and as ftucco for infide work, for the fake of a fine grain and even furface, muft have a greater quantity of lime in its compofition, than is neceffary for cementing the grains of fand together; the incruftation would, by the accefs of acidulous gas after it is laid on, be apt to fwell and chip and lofe the even furface, if the lime were frefh when it is ufed in this

excefiive

exceffive quantity. But this inconvenience is obviated by their proceffes, in which the lime, whether flaked into water or otherwife, imbibes a confiderable quantity of the gas, and is therefore the lefs apt to blifter or fwell, after the ftucco is laid on.

The builders confidering the plaifterers mortar, or ftucco as a finer and better kind of mortar, think it not amifs to imitate them in thofe particulars which are not attended with any expence, and efpecially in the practice of flaking a great deal of lime at once, and of keeping the mortar made fome time; and they do not feem to know, that fuch meafures prevent the mortar from ever acquiring that degree of hardnefs in which the perfection of mortar truly confifts.

SECTION VII.

Of the Depravation of Mortar by the common Method of using the Water; and of the Use of Lime-water.

FINDING by reason and experience, the advantage of totally expelling the gas, and preventing the return of it to lime or even to mortar before it is used; and knowing that common water, which is employed in great quantities, first in flaking lime, and then in making mortar, contains a great deal of the noxious gas; it occured to me that the vulgar process of making mortar is in this fresh instance injudicious, as it tends to injure materials otherwise good.

They flake lime in such a manner that almost the whole of the water is evaporated, and contributes nothing to the mortar, except so far as it deposites its gas in the lime and injures it; and then the flaked dry lime and the sand, require more water to make them into mortar. I have found the quantity of water used for both these purposes, to be twice the weight of the lime, at the least.

The quantities of acidulous gas known to be contained in the waters commonly used in making mortar, must greatly debase the lime which is thus exposed to double its weight of such water; and upon these grounds I was assured, a priori, that it would be a considerable improvement in mortar, to use no water in it except what has been previously freed from acidulous gas.

This is done in making lime-water; and the use of lime water appeared advantageous in another point of view. One seven hundredth part of lime water being lime, according to the experiments of Mr. Brandt which I find to be true; and this lime being introduced in a state of solution which favours the crystalination of it between the grains of sand, assists in cementing them together by the utmost attractive forces of its parts, if my notions of the polarity of these parts be true.

I MADE divers experiments to try the practical validity of this reasoning, and found it to be true: for on comparing specimens of mortar made with my best lime flaked with river water, and sand and water, and spread on tiles soaked in water, with other specimens made with the same proportions of lime flaked with lime water and sand and lime water, and spread on tiles previously soaked in lime water, the latter, at every age of them, were sensibly harder, and they adhered to the tiles better than the former. I must observe however, that such distinctions cannot easily be made, except by those who have a great deal of experience in these trials and comparisons. On repeated examinations of these and my other specimens, I was highly encouraged in my pursuit; for those made with lime water were better near the surface than any I had ever made; and I had good reasons to be persuaded that the extraordinary induration would proceed in time through the whole mass.

SECTION VIII.

Experiments made with a View to approximate the best Proportions of Lime Sand and Water, for Mortar.

IN reading over my notes, and examining the specimens of mortar which I had hitherto made, I perceived that those were the best which being made with common fresh lime, or with well burned lime, contained the least of it; that is one ounce of lime in six or more of sand; and finding this quantity of lime to be much less than is used in the common way of making mortar; and suspecting that as a wall may be the weaker for its containing too much mortar, which widens the joints, so mortar may be weakened by the introduction of more lime than is necessary to cement the grains of sand together; I thought another cause of the defect of common mortar opened to my view; and that it was advisable to determine by experiment, what is the best

propor-

proportion of lime to sand, in making mortar in which lime water is used.

I made five parcels of mortar with my best stone lime recently flaked with lime water, and with the coarse Thames sand, in the following proportions by weight.

1.

Slaked lime — — 1
Sand — — 4
Lime water, q. s.

2.

Slaked lime — — 1
Sand — — 5
Lime water, q. s.

3.

Slaked lime — — 1
Sand — — 6
Lime water, q. s.

4.

Slaked lime — — 1
Sand — — 7
Lime water, q. s.

5. Slaked

5.

Slaked lime — — 1
Sand — — 8
Lime water, q. f.

This latter specimen was not sufficiently plastic for common use; or as the workmen express themselves, it was too short. I further observed that the quantity of water required to make mortar to the proper temper, is greater as the quantity of lime is greater relatively to the quantity of sand.

I spread these on tiles in the month of June, and exposed them to the air and the sun, which then was very hot.

As my former experience taught me to expect that some of these, in hasty drying, would crack considerably; and as mortar, in building, is not liable to dry so quickly as these specimens; in order to render the inferences from these experiments the more general, I made five other parcels of mortar in the same manner and exposed them in the same way, in every respect, except that the direct rays of the sun could not fall on them

or

or heat the pavement on which they stood. In three days I found this necessary, for the first of those which stood exposed to the sun cracked considerably, the second cracked less, the third shewed three or four very slender fissures visible only on a very close inspection, the fourth and fifth shewed no cracks at this time, nor in a month afterwards, when the fissures of the others were considerably enlarged.

Of the specimens kept in the shade and examined on the third day like the former, the first was cracked in divers parts, the second shewed two or three very slender cracks, the rest were not cracked in the least, and never cracked afterwards, although I was forced to remove them to the place where the others stood.

Thus it appeared in a very short time that an excess of lime disposes mortar to crack, and consequently injures it; that the highest proportion of lime to such sand, which may be used without incurring this inconvenience, depends on the circumstances in which the mortar is to be exposed; that no more than

one

one part of lime to seven of coarse sand ought to be used in mortar which is to dry quickly; and less lime may not be used, because it does not render the mass sufficiently plastic for building or incrustation; and that if a greater proportion of lime to such sand improves the mortar in any respect, it is to be used only where the mortar cannot dry so quickly as it did in the specimens exposed to the sun.

In the course of nine months I clearly perceived that those specimens which stood in the shade for the first three days, were harder, and better in other respects, than those which were suddenly exposed to the sun, the comparison being made between the specimens which contained the same proportions of lime, and which cracked the least, or not at all: and of all the specimens, those were the best which contained one part of lime in seven of the sand: for those which contained less lime, and were too short whilst fresh, were more easily cut and broke, and were pervious to water; and those which contained more lime, although they were closer in the grain, did not harden

so

so soon or to so great a degree, even when they escaped cracking by lying in the shade to dry slowly.

I THEREFORE concluded that hasty drying injures mortar made in any proportions of such sand and the best lime; and that the best proportion is one of lime in seven of sand, whether the mortar is to be quickly dried or not.

I MUST observe however that these conclusions were made rather with a view to my future experiments, in which an approximation to the best proportions of lime and sand and the best treatment of the mortar would save a great deal of trouble, than to any general and invariable rule for making mortar.

I RESERVED it to be mentioned in this place, that I set apart four ounces of each of the foregoing specimens of mortar, and spread these portions severally on plates of thin window glass, to the thickness of a quarter of an inch or thereabouts; and I noted the weight of each plate with its specimen of mortar recently made.

THESE

THESE being equally expofed to the fun and weighed at different periods were found to lofe weight in equal times nearly in the proportion of the quantity of lime or of water ufed in making them; and the fmalleft lofs of weight when the fpecimens were perfectly dry and confiderably hardened, was one-tenth of the weight of the fame fpecimens recently made.

IN many former experiments I had obferved, but referved it to be mentioned in this place, that mortar which fets without cracking, whether this be owing to the due proportion of fand, or to the flow exhalation of the water from mortar containing lefs fand; never cracks afterwards, whatever other faults it may have: the fpecimens mentioned in this fection, after a trial of eighteen months afforded the fame obfervation.

BY the fetting of mortar, I underftand that folidity which it acquires by mere drying, and which differs widely from the induration that takes place in time by other means which we fhall prefently confider.

SEEING

SEEING then that the quantity of water in mortar is as the quantity of lime, that the fissures happen only in the drying, or setting, that the danger of cracking is greater, not merely as the quantity of water is greater relattively to the sand, nor merely as the water is more expeditiously exhaled, but in a rate compounded of these; I inferred that mortar which is to be used where it must dry quickly, ought to be made as stiff as the purpose will admit, that is, with the smallest practicable quantity of water; and that mortar will not crack, although the lime be used in excessive quantity, provided it be made stiffer or to a thicker consistence than mortar usually is.

THIS inference was afterwards found to be true: for specimens made thus with one part of lime and only six of the sand, and others made with greater proportions of lime, but as stiff as they could be used, did not crack, in any exposure; but they had faults which will be hereafter noticed.

SECTION IX.

Theory of the Induration dependent on the Proportions of Lime and Sand in Mortar, and Observations on the bad Effects of the vulgar Proportions of these.

IT is sufficiently known that the aggregation of calcareous bodies, which burn to lime, or are chiefly composed of the matter of lime, is much weaker than that of the quartose; insomuch that the steel which easily cuts all calcareous stones or spars, is as easily cut by the siliceous; and all stones, or powders which are chosen for cutting or grinding steel, are found to have this effect by reason of their siliceous or quartose particles.

THIS being considered together with divers observations heretofore related, I reasoned in the subsequent manner.

As stones are cemented together in walls, by the mediation of mortar, so the grains of
sand

sand or gravel are made to cohere and form a solid mass of mortar, by the intervention of lime.

By the bare inspection, as well as by the experienced induration, one part of lime paste appears sufficient to intercede the grains of seven of sand without interruption of continuity, and in drying to fill the spaces between them, or to attract matter enough for this purpose from the air.

In this case the grains cohere at the smallest distances of them, and by means of the thinnest laminæ of calcareous matter; and such mortar is the stronger as it consists of the greater quantity of hard quartose bodies cohering by means of the smallest practicable quantity of softer and brittle calcareous strata; just in the same manner as a wall, built with porphyry and bad mortar, is, cæteris paribus, the stronger as the joints are made thinner: for all masses of such structure as mortar or cementious walls, resist fracture and ruin, with powers of aggregation which are, not merely as the aggregation of the stones or bricks, nor barely as the aggregation of the softer cement, but in a ratio compounded of these,

these, and varying with circumstances which we need not attend to at present; and those masses therefore will resist the most, in which the stronger aggregates bear the greatest proportion to the weaker, so far as it is consistent with the continuity of them.

SECONDLY, The small stones which compose a heap of sand do not imbibe water: their volume is not encreased by wetting them nor lessened in drying; neither does a measure of wet sand contract sensibly in drying: this last I have repeatedly experienced. But small bits of lime are considerably increased in bulk by wetting them; and as the soft paste of lime contracts greatly in drying, it must crack in every part where the drying paste is prevented, by its adhesion to bodies or by other causes, from contracting uniformly and concentrically. As the contraction of mortar in drying, and its consequent cracking, depend on the lime paste, and not on the sand, they must take place in the greater degree, as the quantity of lime and water is greater, and they must be lessened or prevented by a due proportion of sand; which proportion experience shews to

be

be feven parts of fand to one of lime. Thus we underſtand the cauſes of cracking, and how it happens that this defect is prevented by uſing leſs than the cuſtomary quantity of lime; and, although the lime ſhould be uſed in exceſs, by uſing leſs than the uſual quantity of water,

THIRDLY, The more perfect and expeditious ſetting and induration of mortar containing only one part of lime in ſeven of fand, than of mortar made with greater proportions of lime, may be deduced from ſeveral concurrent cauſes. Having leſs water in its compoſition, it dries ſooner, and the calcarious matter cryſtalizing more quickly in it, gives it ſooner that ſolidity which we expreſs by the word ſetting. Having leſs lime in its compoſition, it is ſooner faturated with the matter which the air preſents, and which ſeems neceſſary to the induration of mortar; and in this faturation, the ſwelling of the lime is not ſo great as to puſh forth and derange the grains of fand after they have once been placed and in ſome degree cemented together.

THIS

This latter inconvenience arifing from the excefs of lime, cannot eafily happen in mortar compreffed on all fides in maffive buildings; but it manifeftly occurs in the exterior parts of the joints in walls, where the mortar vifibly fwells and after fwelling crumbles: it is likewife vifible in the upper part of walls of modern conftruction, where the fwelling is not prevented by a fuperincumbent weight. In thefe cafes the joints become hollow; houfes lately built look old or ruinous; and the bricks themfelves, being bibulous, in fuch expofure foften and moulder, in confequence of the alternate wetting drying freezing and thawing; thefe being effectual agents in the diffolution of all bodies which freely imbibe moifture.

Without awaiting the event of thofe experiments which I have lately made on the great fcale, and fhall point out before I conclude this effay, we may on thefe grounds alone affure ourfelves that the ftrength and duration of the calcareous incruftations compofed of lime and fand, will be greater, as we depart further from the proportions of lime and fand commonly ufed, approaching

to

to that of one part of lime to seven of sand; because the stucco which hardens the soonest must be the least injured, whilst it is new, by the beating rains, and various accidental impressions; because that which adheres most firmly to the other materials of buildings, and which acquires the greatest degree of induration, must contribute most to the strength of the walls, and best withstand the shocks, attrition, and other trials to which the stucco is exposed; because that which contains the greatest proportion of sand, is less liable to be injured by any saline matter with which the air is sometimes impregnated, as its calcareous matter is the better defended by the sand: but above all, because the stucco made with one part of lime and about seven of sand, is not disposed to crack: for incrustations in this climate perish sooner by reason of the fissures than of any other defect; because the water imbibed into the slenderest of them, as well as into those which appear on a cursory view, swells in the congelation, and dilates them; and frequent alternations of wetting and freezing, gradually widen them, until the stucco is bulged and torn from the walls.

SECTION

SECTION X.

Experiments on old Cements, authorizing the Proportion lately recommended of Lime and Sand.

TO difcover the quantity of lime and fand originally ufed in any hard and old cement which I find by a previous analyfis to confift of lime and fand or clean gravel, I break a pound of it into fmall fragments, but not into powder, and with diluted marine acid I diffolve and wafh away the calcareous part from the gravel or fand. I meafure the acidulous gas obtainable during the folution; and knowing the weight of any quantity of it in any temperature or weight of the aerial atmofphere, I fubtract this weight of acidulous gas and that of the fand or gravel, from the whole weight of the mortar, and ftate the refiduary weight as that of the lime originally employed; knowing that it could not have made fo hard a cement, if it had not been fo far burned as to retain very little acidulous gas. I did not adopt this method of examination before I had found it to exhibit the

the lime and sand of my oldest and hardest cements in the same proportions in which I had mixed them.

By this kind of analysis, and by other trials, I found that the quantity of lime in old cements made with clean sharp sand, and noted for their hardness, was much less than is now commonly used in mortar; and that in the hardest, it was very near to that which my experiments indicate to be the best.

By sharp sand I mean that whose grains are bounded by flat surfaces.

Thus I found the inferences made from my compositions to be authenticated by long experience, so far as they relate to the proportions of lime and such sand.

SECTION

SECTION XI.

Experiments and Observations shewing the Agency of acidulous Gas in the Induration of Mortar, and Circumstances which impede or promote it. Practical Inferences.

THE observations made on divers specimens of mortar, at different periods, led me early into the opinion that the setting of mortar depends chiefly on the exsiccation of it, but that the induration is principally owing to the accession of acidulous gas in certain circumstances, and not to the drying, as the workmen generally imagine. In order to place this opinion beyond a doubt, and to discover the circumstances which favour or impede this induration, I made the subsequent inquiries.

Experiment I.

I MADE mortar with seven parts of the Thames sand, one of the best flaked chalk lime and the necessary quantity of lime water, and forming a part of it into oval pieces,

I put these into a gallon bottle, which I stopped closely with a ground glass stopple waxed; and I noted the gross weight of the bottle and mortar, and placed it exposed to the sun. Having examined it frequently during the first month, I could perceive no alteration in the weight, nor any thing worth notice, except that some water exhaled out of the mortar, and condensed in bright drops on the sides and upper parts of the bottle. At divers times during six months afterwards I shook and weighed the bottle, and found the mortar quite soft and the weight of the whole unaltered.

EXPERIMENT 2.

ANOTHER portion of the same mortar was spread briskly, as soon as it was made, on oblong pieces of dry and warm tile, and these were immediately placed over a sand bath, where they were gradually heated to about an hundred of Farenheit for six hours, and then to an hundred and fifty for two hours more, when the mortar was dried thoroughly. I took particular notice of the solidity which it acquired in this hasty drying, and then put the pieces of tile with the adhering

hering mortar into a bottle closely stopped in the manner already described, marking the gross weight.

At the expiration of seven months, I found the whole unaltered in weight, and the mortar as easily cut or broken as it was when I put it into the bottle.

Experiment 3.

Another part of the same mortar was spread whilst fresh on a large tile, to the thickness of half an inch, and the tile was immediately placed in a tub, in which I put water to the depth of three inches over the mortar, and which I placed in the open air to receive the rain. At different times I broke the calcareous pellicle which formed on the water and defended it from the air, during the first fortnight: afterwards the wind and rain rendered this precaution unnecessary. In the course of six months, the mortar, instead of acquiring any solidity, was deprived of the greater part of its lime; and what remained on the tile, was not much different from a layer of wet sand.

Experiment 4.

Another portion of the same fresh mortar was spread on a board strewed with flaked lime to prevent adhesion, and placed in the open air, but sheltered from the sun. When this mortar became sufficiently solid, which was on the second day, I raised the pieces, which were about a quarter of an inch thick, from the board, and set them upright, fully exposed to the weather, which was about this time dry and warm. In seven weeks after this exposure they were indurated to a considerable degree. They resisted a cutting instrument nearly as much as Portland stone does, but not so well any force tending to break them across at once. I then placed them under water as I had done by the former portion of this mortar.

After they had lain in the water four months I examined them attentively and found them, if at all altered, to be rather softened than indurated further. I replaced them in the water, to be better satisfied about them; but by mistake, they were removed in my absence, and lost.

EXPERIMENT 5.

Soon after the foregoing parcel of mortar was made, I prepared another in the fame manner, and fpread a part of it on a tile foaked in lime water, and placed the tile in the open air fheltered from the fun and rain. After it had ftood a month in this fituation, I placed it where it was fheltered only from the rain.

EXPERIMENT 6.

Another portion of the fame mortar was fpread frefh on a warm dry tile, which I placed over a fand bath, where the mortar was heated to about an hundred of Farenheit for fix hours and then to an hundred and fifty for four hours more, at the expiration of which it was folid and perfectly dry. The next day I placed it in the open air, expofed to the fun, and the weather, which was dry and warm for a confiderable time afterwards.

On comparing thefe two laft at the expiration of feven months, and again after fix months more, I could eafily perceive that
the

the latter was inferior to the former; for it was much more eafily cut and fcaled from the tile and broken.

Experiment 7.

WITH mortar made, the day after I had made the former, of the fame materials and in the fame manner; and with new bricks, which I had heated almoft to rednefs and fuffered to cool to the temperature of my hands; I brifkly erected a little wall half a brick thick, on a ftone bench raifed for the purpofe and fully expofed to the weather.

Experiment 8.

ON the fame day and with the like mortar, and with cold new bricks previoufly foaked in lime water, I erected another wall, equal to the former in dimenfions, and placed in the fame manner on a ftone bench in the open air.

AFTER nine months, in pulling thefe walls to pieces and in divers comparifons of the cement of them, I found that the latter cement adhered better to the bricks and was

harder

harder than the former, infomuch, that I had not a doubt about it.

EXPERIMENT 9.

IN a few days after I had made the experiment with the warm bricks, I confidered the walls erected in variable weather, and the fence walls which are wetted frequently and deeply whilft new, by rain, or by moifture from the ground, and as often dried quickly; and I was defirous to learn the effect of fuch alterations of wetting and drying.

I THEREFORE fpread mortar, made like thofe parcels lately mentioned, on a large tile foaked in lime water, and as often as it had dried, in fair weather, and generally at the interval of three days, I wetted it with rain water. In the courfe of nine months I found it was much lefs indurated than the fpecimen made in the fame manner, and defended from the rain: it moreover grew green by means of a vegetation which took place on the furface of it, and which thrived the more as the mortar was frequently wetted, or the tile longer fuffered to lie flat on the ftone bench already mentioned.

I HAD often observed such a vegetation on mortar which I had made a few months before, especially when, in the summer season, I laid the tiles flat on the wooden border of a dust-hole, or when, for want of room to preserve the specimens in, I piled many of them together in a damp corner on the pavement: I likewise saw that where the vegetation took place, the induration did not proceed as it does elsewhere: on the contrary, semi-indurated mortar softened there.

ALL these being considered, I was satisfied that frequent wetting or constant moisture, together with exposure to air, injure mortar in a great degree, if it be not perfectly indurated by great age before it is exposed to such trials; and that the vegetation depends chiefly on moisture.

EXPERIMENT 10.
BY the kind of analysis mentioned in the tenth section, I repeatedly examined the proportion of acidulous gas to the lime, in the hardest of the old cements which I had collected; and finding it in the best of them to be, at the lowest, in the proportion of three

to five, I rate the quantity of acidulous gas imbibed by good mortar, during the induration of it, at sixty pounds at least, for every hundred pounds of lime.

EXPERIMENT II.

Such mortar as that of the first experiment of this section was formed into slender pieces, each an inch broad, a quarter of an inch thick, and three inches in length. These were placed in an airy passage, sheltered from the sun and rain, and were turned as soon as they could bear it without danger of cracking; they were then set upright and fully exposed on all sides to the air.

On the fourth day I slid four of the pieces entire into a small wide-necked glass retort, which I set deep in a sand bath, with its nozzle immersed in quicksilver, which stood cool whilst the charge was gradually heated, in the course of forty-eight hours, to about seventy-five of Farenheit, which is under the temperature of incrustations of this kind exposed to the sun in summer; and in the course of forty eight-hours more was slowly heated to about an hundred of Farenheit, to

which

which degree incruſtations are frequently heated by the ſun in ſummer. As the retort cooled I admitted the neceſſary quantity of air, and then left it, with the nozzle immerſed deeply in the mercury, during three months. I then ſlid the pieces gently out of the retort, after having wiped away a few drops of water which adhered to the veſſel in their way; and immediately made the compariſon which I ſhall preſently mention.

CLOSE to the retort, and in a ſituation where the heat was equal to that deſcribed, or nearly ſo, I placed four other of the pieces above deſcribed, on the fourth day after they were made: I encompaſſed them with the ſand, but ſecured a free acceſs and even a circulation of air to them. When the ſand bath was cooled, I put theſe pieces, which were thus perfectly dried, into a bottle which I ſtopped cloſely in the manner heretofore deſcribed.

ON the ſeventh day after the pieces were made, on the twenty-firſt, and at the expiration of three months, I examined four pieces taken from different quarters of the remaining

remaining parcel, and found the quantity of acidulous gas which they yielded, to correfpond with the degree of induration and the depth to which it had advanced in them refpectively.

On comparing and examining the pieces dried in the retort and kept three months in it; the pieces dried in the fame heat and freely expofed to air during four days, but afterwards kept in a clofe veffel; and the pieces which dried and hardened in the free air, without being heated; I found that the firft were friable in comparifon with the fecond, and the laft were by much the hardeft and beft,

As the fecond tenth and eleventh experiments, together with obfervations formerly made, fhew that the induration peculiar to mortar, is not caufed by exficcation; that it is greater, as the calcareous matter of cements approaches nearer to be faturated with acidulous gas; that it is retarded or prevented, as the acceffion of acidulous gas is interrupted or obviated; we may conclude that this matter is a principal agent in the induration of
calca-

calcareous cements and indispensibly necessary to it.

By observations formerly made, but especially by the comparison of the fifth and eight experiments of this section with the sixth and seventh, I learned that hasty drying prevents good mortar from ever acquiring the hardness which it otherwise would have; and that the more slowly the proper water of the mortar is exhaled or absorbed from it, in incrustations or brickwork, the more perfect will be the induration of it.

By the first third and ninth experiments of this section compared with the fourth, fifth and others, and by observations which led me to make these experiments; I discovered that mortar which is not suffered to dry, or which is supplied with moisture as fast as its proper water exhales, does not harden, or hardens only to a small degree by any accession of acidulous gas.

The fourth experiment indicates that mortar, whose lime has not yet imbibed its complement

plement of acidulous gas, although the mass be considerably hardened, is liable to be injured by soaking in water, if it be pervious to water so freely as these thin pieces were.

ALL these experiments and observations conspire to point out the circumstances in which mortar becomes indurated the soonest and in the highest degree, and operates most effectually as a cement. To this end it must be suffered to dry gently and set; the exsiccation must be effected by temperate air and not accelerated by the heat of the sun or fire: It must not be wetted soon after it sets; and afterwards it ought to be protected from wet as much as possible, until it is compleatly indurated: the entry of acidulous gas must be prevented as much as possible, until the mortar is finally placed and quiescent: and then it must be as freely exposed to the open air as the work will admit, in order to supply acidulous gas, and enable it sooner to sustain the trials to which mortar is exposed in cementious buildings and incrustations.

FROM these considerations we learn other causes, besides those already mentioned, of the speedy ruin of our modern buildings.

THE

The mortar made with bad lime and a great excefs of it, and debafed in watering and long expofure, is ufed with dry bricks and not unfrequently with warm ones. Thefe immediately imbibe or diffipate the water and not only induce the defect above noticed, but, as the cement approaches nearer to be dry, whilft it is ftill liable to be difturbed by the percuffions of the workmen, render it more nearly equivalent to a mixture of fand and powdered chalk.

But to make ftrong work, the bricks ought to be foaked in lime water, and freed from the duft, which in common bricklaying, intercedes the brick and mortar in many parts. By this method the bricks would be rendered clofer and harder; the cement, by fetting flowly, would admit the motion which the bricks receive when the workman dreffes them, without being impaired; and it would adhere and indurate more perfectly: the fame advantages would attend the foaking of bibulous ftones in lime water, and the ufe of grout; provided this were made with good lime fand and lime water.

In plaiftering, the workmen always brufh away the duft and wet the wall on which they are to lay the cement, becaufe it will not otherwife adhere. From what has been already faid it is manifeft that this ought to be done with lime water, and repeated as long as the wall is thirfty.

To perceive more clearly how much our flight buildings are weakened by the agitations and percuffions to which they are expofed, firft in erecting the walls and fettling the timbers, and then in driving thofe wedges to which they faften the wainfcots cornifhes and other ornaments, we muft obferve that the acceffion of acidulous gas to mortar, was found to contribute nothing to the ftrength of it, when it entered the compofition before it was finally fixed in a quiefcent ftate: and a little experience is fufficient to teach us, that the fame matter which affifts in the induration of mortar, never ferves to repair the fiffures, or folution of continuity between the bricks and cement, which happen after it is fet. When mortar is fet, and before it is indurated, it may eafily be ferved from the
bricks

bricks and crumbled; and for want of softnefs it cannot bend into the fiffures, or refume its former condition in any time. Therefore by heavy blows, and in wedging, our walls muft be greatly weakened; and the more, as the houfes are flight, quickly built, and haftily finifhed.

SECTION

SECTION XII.

Experiments shewing the best Kinds and Mixtures of Sand, and the best Method of using the Lime Water, in making Mortar.

PURSUING the analogy intimated in the ninth section, I thought that as large stones with carvilinear faces, bedded in common mortar, do not form so strong a wall as they may when their interstices are filled with fitting stones together with the due quantity of mortar; so mortar made with sand, whose grains come near to be equal in size and globular, cannot be so strong at any period of induration, as that which is made with the same mixed with as much fine sand as can easily be received in its interstices, in order that the lime may cement the grains by the greater number and extent of contiguous surfaces. By this notion I was excited to make provision for a new series of experiments.

I CLEANSED a large quantity of the Thames sand, by washing it in streaming water, and

sorted

sorted it into three parcels: the coarsest, which I call the rubble, consisted of small pebbles, fragments of weathered shells, and grains of sand of divers sizes, which in washing had passed through a sieve whose apertures were one eighth of an inch square, but could not pass through a brass wired sieve, whose meshes were one sixteenth of an inch square, or rather larger: the next parcel, which I called fine sand, consisted of grains of divers sizes, which in washing passed through a finer sieve whose meshes were one thirty-second of an inch square: the third parcel consisted of grains the largest of which were washed through the coarsest sieve, and the smallest were retained, in washing, on the fine sieve: this I call coarse sand.

It is to be observed that the sand which can pass through a sieve, in washing, is considerably finer than that which may be sifted through the same sieve, when it is dry.

Having dried these parcels on a sand plate, and provided a narrow mouthed glass bottle capable of holding about two ounces troy of water, and a cylindrical glass vessel which contained

tained twelve of these measures, I found by repeated trials, that the large vessel, charged to the brim with my rubble, might be made to hold somewhat more than one additional measure of it, when the rubble was well packed, by striking the bottom of the vessel frequently against the table perpendicularly. Charging the same vessel with coarse sand, I could by the same treatment make it hold two thirds of the thirteenth measure: and twelve measures of fine sand were so far contracted in this motion of the vessel, that it could hold one measure and one fourth more, or thirteen and one fourth in all.

After noting how far the intersticial spaces in each sized sand can be lessened by packing, I used water to shew what proportion these bear to the solids, in these different sands. I found that the thirteen measures of rubble which I stowed into the glass cylinder, could take in five measures of water, without any increase of bulk; or rather with a striking decrease of bulk: the twelve measures and two thirds of stowed coarse sand, imbibed four and one half of water, and yet decreased sensibly in bulk: the thirteen measures

measures and one fourth of fine sand packed, could drink in only four measures of water; but the diminution of bulk was more considerable than in either of the former; for the sand and water together measured less by one seventieth than the packed sand alone.

When sand was poured into the glass cylinder until it was filled, and water was added before the sand was packed, by a slight agitation of the vessel the sand contracted in a much greater degree than is above expressed. Upon the whole it seemed that water, by poising the grains, facilitates their sliding on each other to fit well and fill the spaces.

Until I had made these experiments I did not well understand, how the beating of new mortar makes it much wetter, and more plastic withal, than it can be made with the same proportions of water and solids, by mere mixture. I now perceive that beating produces this effect by closing the interstices of the sand, and rendering a small quantity of lime paste as effectual towards filling them and holding the grains together to form a plaſtic maſs, as a greater quantity is, in ſand whoſe grains cannot fit each other ſo well.

Seeing that the interſticial ſpaces in ſand are ſo greatly leſſened by wetting it, I judged it expedient, for this reaſon alone, to expend all the water I ſhould henceforward uſe in making mortar, in wetting the ſand compleatly. I afterwards obſerved other advantages ariſing from this practice: for in filling the ſpaces with the fluid, the air is eaſily expelled, and the lime equably diffuſed in them by a little beating: but when the water is added to a mixture of lime powder and ſand, the air is entangled in the lime paſte, and cannot, without a great deal of beating, be totally preſſed out of the plaſtic maſs: I likewiſe found, that as an exceſs of water is injurious in mortar, this is an excellent method of regulating the quantity of water; for the portion of lime water which fills the ſpaces in ſand, and can be held by capillary attraction in a flat heap of it, is preciſely the quantity which makes well tempered mortar with one part of the beſt flaked lime and ſeven of the beſt ſand.

As I found some difficulty in expelling the air bubbles out of the sand wetted in my deep cylindrical measure, even when I stirred up the mass with a slender instrument, I concluded that the spaces in sand are rather in a higher proportion to the solid substance of it, than they appeared in these trials: so that we may say they are at least one third and more of any measure of the fine sand, greater in coarse sand, and greatest in the rubble. Suspecting on another ground that these experiments did not shew the whole of the spaces in sand, because water tends to insinuate itself between the contiguous faces of the grains, and consequently to remove them asunder, even whilst it arranges them; I attempted to ascertain the proportion of these spaces to the solids, by another method founded on this supposition, that the measured portion of sand which weighs the most, has the smallest quantity of intersticial space.

By the experiment, I found that a well packed measure of the rubble weighed twenty ounces three pennyweights: the like measure of the coarse sand packed, weighed twenty one ounces eighteen pennyweights:

and the same quantity, by measure, of the fine sand, weighed twenty three ounces two pennyweights and twelve grains.

This trial corresponds sufficiently with the former in shewing, that the sum of the spaces in the rubble, is much greater than that of the coarse sand, and that the spaces in this are larger in the sum, than those of fine sand.

In order to learn whether this proportion is maintained in all kinds of sand, I tried by water and by weight in the foregoing manner, a great number of sands used in London; such as the coarsest glass-grinder's sand, Hampstead-sand, Lynn-sand, fine house-sand &c.

The result of these experiments taught me that the spaces are always smaller as the sand is finer, provided the comparison be made between the sorted fine part and the coarsest part of any kind of sand: but this does not hold true in the comparison of fine sand and coarse sand of different districts.

On examining the several specimens of sand with a lens, I perceived that, in some, the grains, however different in figure, were bounded by flat faces meeting each other in angles; whilst in others, the faces were generally rounded, and the figure such as the foregoing grains would be reduced to by grinding off their angles. The first kind I call sharp sand, the other round sand. Then taking into consideration the measurement already described, together with the sharpness or roundness of the sand, I found that the spaces are, in different kinds of sand, as the size and roundness of them compounded; but they do not appear to be smaller in any kind of sand that I have seen, than in our fine parcel of Thames sand; which I think is owing to its being sharper than any of the finer sands which I had compared it with. The measure which contained twenty-three ounces two pennyweights twelve grains of the fine Thames sand, contained only twenty-two ounces ten pennyweights of the Lynn sand, which is a great deal finer, but rounder.

Having thus found the kind of sand which, by reason of the size and figure of the grains, has the smallest intersticial space; I next endeavoured to ascertain the mixture of coarse and fine sand, which lessens this space in the greatest degree; which therefore requires the less lime to cement the grains together, and for the reasons already mentioned, promises to make the hardest and most durable cement.

I FOUND that nine measures of the shingle, and nine measures of the fine sand, both well packed, measured when mixed and stowed closely sixteen measures and one-eighth: that eighteen measures of the shingle and nine of the fine sand tried in the same way, measured twenty-four: and that on mixing the shingle and fine sand in various proportions, nine measures of shingle took, into its interstices, one measure and one half of the fine sand, without any increase of bulk.

I NEXT learned that nine measures of the coarse sand and nine of the fine, measured in the like manner seventeen and a half: that eighteen such measures of coarse sand well mixed with nine of the fine sand, measured twenty-six: and that on mixing these sands

in

in various proportions, eighteen meafures of the coarfe fand took into its interftices one meafure of the fine fand, without any increafe of bulk.

Lastly I found that eighteen fuch meafures of the coarfe fand and nine of Lynn fand, which is much finer grained than the foregoing, meafured twenty-four when well mixed and ftowed: and that on mixing them in various other proportions, nine meafures of the coarfe fand took into its interftices one and a half of the Lynn fand.

By thefe and a variety of fimilar experiments made on different fands, I found that the quantity of fine fand taken into the interftices of a coarfe fand, was the greater without increafe of bulk, as the grains of the coarfe differed more from thofe of the fine in bulk, provided the diameters of the grains of coarfe fand did not in general exceed thofe of the fine, in a proportion greater than five to one; that the greateft quantity of fine fand, which could be taken into the interftices of coarfe fand, was one-fixth of the bulk of the coarfe fand; and that in general the

mixture

mixture of fix meafures of coarfe fand with one of the fineft fand, reduced the fum of the interfticial fpaces to nearly one half of the quantity of them in coarfe only, or in fine Thames fand or rubble only.

INSTRUCTED by thefe obfervations I proceeded to the following experiments, in order to learn the advantages or defects attending each kind of fand, and how far my expectations from the art of leffening the fpaces, were well founded.

I MADE feveral parcels of mortar with my chalk lime lime water and rubble in different proportions; the quantity of lime being in one a fourth of that of the rubble, in another only one-feventh, and in the others intermediate: I made other parcels of mortar with my chalk lime, lime water, and the coarfe fand; and others with this lime, lime-water and the fine Thames fand, in the laft mentioned proportions.

I NEXT made a great variety of fpecimens of mortar; fome of which confifted of rubble and coarfe fand mixed in different proportions,

portions, wetted with lime-water, and blended with one-fourth or one-seventh or intermediate quantities of lime: others were composed of similar mixtures of rubble and fine sand with lime and lime water; and others consisted of rubble coarse sand and fine sand mixed in different proportions, wetted with lime water, and beat up with the different quantities of lime lately mentioned.

I SPREAD a part of each of these specimens of mortar, as soon as it was made, on a tile soaked in lime water, half an inch thick in some places, and much thinner in others: I placed the remainder of it, formed into oblong pieces of about an inch diameter, on the part of the tile which was not covered with mortar; and I set all the tiles numerically marked, in the situation formerly desc ibed, where they were equally exposed to the weather: this was done in May, 1777: during the succeeding twelve months I examined each specimen, and noted my observations, the most useful of which I shall endeavour to relate in a few words.

THE

The specimens containing rubble and lime mixed in any proportion greater than five to one, were not fat enough, when fresh, to be conveniently used in building or stuccoing: but none of them, not excepting those which contained the greater quantities of lime, cracked in drying. Those which had the smaller quantities of lime in them, were very rough on the surface, coarse in the grain, spongy, and easily broken: they shewed a defect of lime, because those which contained more lime were not so bad in these respects. By all of them it appeared that whenever such rubble must be used, for want of sand or finer gravel, the lime mixed with it must not be less than one-fifth of the quantity of rubble.

Of the specimens consisting of coarse sand and lime, those which had the smaller quantities of lime were too short for common use, and could not be made to assume a close and smooth surface, whilst fresh; but in drying and hardening, they were in every respect preferable to the cements made with rubble and lime, in the same proportions: and of the same specimens, those were the best which
contained

contained one part of lime in five of sand; the others containing less lime being faulty like those made with rubble, and those in which the lime was mixed in much greater quantity, having the faults often observed to attend the excess of lime.

The specimens which consisted of fine sand and lime, were in general better than the foregoing: and that particularly which contained one of lime in six and an half of sand, was in all respects much better than those made with the same or any other quantities of rubble and lime, or coarse sand and lime. The specimen which was formed with seven parts of fine sand, and one of lime was not so compact and hard as that last mentioned. The comparison of these two shewed that seven of sand are too much for one part of lime, when the sand is fine and unmixed with coarse grains. The specimen made with four parts of fine sand, and one of lime had the noted faults attending the excess of lime; for it cracked in drying, and was sensibly injured in the winter, by those alternations of drying, wetting, freezing, and thawing, formerly noticed.

On

On divers comparisons of those portion of mortar made of fine sand and lime, with the former, I was persuaded that a better cement can be composed with such sand as I call fine, than with a coarser sand, whose grains are all larger than any of those in my fine sand; provided the coarser sand be not much sharper than all that I have yet seen. If my experiments had been made in slow succession, this last observation would have led me to imagine that mortar will be found the better as the sand is finer.

Of the observations made on the parcels of mortar consisting of mixed sands and lime, those which follow are the most pertinent to our present enquiry.

The specimens made with mixtures of rubble, coarse sand and different quantities of lime, resembled those made with rubble and lime in similar proportions, when the rubble was predominant; and resembled those made with coarse sand and lime, in similar proportions, when the coarse sand was predominant; and I could perceive
no

no advantage derived from the mixture of rubble and coarſe ſand, except that the cement was ſomewhat better as the quantity of rubble was leſs, relatively to the quantity of ſand and lime: but none of theſe ſpecimens were in any reſpect ſo good as thoſe made with fine ſand only.

OF the ſpecimens made with rubble and fine ſand, that was the beſt in which the fine ſand was twice the quantity of the rubble. But I could not perceive that any of theſe ſpecimens were preferable to thoſe made with the like quantities of fine ſand and lime; or that any conſiderable advantage is gained by the mixture of rubble and fine ſand.

Of the ſpecimens made of coarſe ſand fine ſand and lime, thoſe were manifeſtly the beſt, which conſiſted of four parts of coarſe ſand, three of fine, and one part or a little more of lime: for, whilſt freſh, they were more plaſtic than the others, and were eaſily made to aſſume a ſmooth ſurface; they were not diſpoſed to crack in this method of drying; they were not at all injured by wet or freezing or thawing; they were pretty cloſe in the grain; and they

they grew so hard, in the course of nine or ten months, as to resist the chizel, or any force tending to break the oblong pieces, much more powerfully than any of the specimens lately mentioned. I noted them as the best specimens of mortar that I had ever made; and one part of lime, in four of coarse, and three of fine sand, to be a better proportion than any other of the sands and lime, for incrustations.

Of the various specimens of mortar made with mixtures of the rubble, coarse sand and fine sand, those were the best, in which the fine sand was equal or nearly equal in quantity to the rubble and coarse sand; in which the rubble was not much more than one seventh part of the quantity of both sands; and in which the weight of the lime was one seventh of the weight of the sand and rubble, or a little more: But these specimens, when fresh, were less plastic, and less capable of assuming a smooth surface under the trowel, as the quantity of rubble was greater; and I could not find they were preferable in any particular to those respectively which were made with similar quantities of lime and the mixtures of coarse and fine sand lately commended.

Upon

Upon the strictest comparison, I concluded, that one part of rubble in three of coarse and three of fine sand, makes as good mortar with lime, as can be made with the sand and lime without rubble, for any purpose which does not require a finer cement; but there is no advantage gained by the use of rubble where the coarse and fine sand can be had equally cheap, unless a rough surface be required.

In stuccoing walls the rubble promised to be useful in pointing and in the first coat; because a roughness of this coat makes the finer exterior coat adhere more firmly.

In the review of all these specimens it appeared, that the quantity of lime, which forms a mass somewhat plastic with sand and water, is the smallest quantity necessary for making the best mortar which such sand can afford; and that any further quantity of lime is useless in the coarser sands, and injurious in the finer: that the necessary plasticity is induced by the smaller quantities of lime, as the interstices of the sand are smaller in the sum, and as the grains fit each other the better in consequence of the due mixture of coarse and fine

fine fands: but that the leffening of the interfticial fpaces, by the mixture of fine fand with the coarfe, does not enable us to leffen the quantity of lime fo far as might be expected in confequence of our notions of the fpaces meafured by water. It feems that the grains of fine fand are held afunder by the lime pafte, to a greater diftance than they are by water; and that the reafon, why the finer fand requires more lime than the coarfer and mixed fand, is, that the fpaces, which are more numerous in fine fand than in the coarfe, are more augmented in the whole quantity of them, by the particles of lime, which intercede alike the coarfe and the fine grains.

SECTION

SECTION XIII.

Experiments shewing the Effects of finest Sand and quartose Powder, in Mortar: Observations on the finest calcareous cements: Practical Precepts.

THE last mentioned notion led me to suspect, soon after the foregoing experiments were made, that, although the fine Thames sand made better mortar than the coarse sand or the rubble afforded, the mortar will not always be the better as the sand is finer, however sharp it be. I therefore procured a large quantity of the very fine pit-sand used in London under the name of house-sand: I washed away the clay with which it abounds, and dried it: viewing it, when thus cleansed, with a lens, I estimated the size of the grains to be, at the medium of the largest and smallest, about one ninth part of that of my fine Thames sand: this I call finest sand. At the same time I was favoured, by my neighbour Mr. Bentley the ingenious manufacturer of the ornamental Staffordshire ware, with the necessary quantity of the fine powder of calcined flints, which is prepared for his manufactory.

With divers mixtures of these with lime water and lime, in a variety of proportions; and with each and both of these, blended with coarse and fine sand, lime and lime water in similar proportions; I made a great number of specimens of mortar, which I tried in the manner already described; and noting my observations on them, I found the following to be the most eligible for the concise recital intended in this essay.

Mortar containing the quantity of lime necessary to the plasticity and other desireable properties of it, or a greater quantity of lime, is the more liable to crack in drying, as the sand of it is finer.

Mortar made with this finest sand and lime, does not grow so hard, or resist fracture so forcibly, as that made with my fine Thames sand and lime, in the same proportions, or any others nearest to these. But the former mortar, when composed of about six parts

parts of sand, one of lime, and the necessary quantity of lime water, and slowly dried, becomes much harder than any of the common calcareous stuccos, used by plaisterers.

Mortar composed of lime, my fine Thames sand, and the finest sand, is the worse as the quantity of finest sand is greater: and this holds true in every tried proportion of the sands and lime.

Mortar consisting of lime, coarse Thames sand, fine Thames sand, and finest sand, is the worse as the quantity of this last is the greater, when the comparison is made between it and the cement made with the same quantities of lime and the best mixture of coarse and fine Thames sand.

Mortar made with flint-powder, lime and lime water, in any proportion, is more liable to crack in drying, than mortar composed of any sand and lime: it is moreover incapable of hardening to so great a degree; whether the hardness be tried with a chizel, or by breaking it across. But mortar made with about five parts of flint-powder, one of lime, and the neceſſary quantity of lime water, is neverthelefs preferable to any ſtucco now uſed in inſide work, for the finiſhing coat; becauſe it has a more lively whiteneſs, and aſſumes a finer furface, which I think might be made to imitate that of marble: It requires however to be dried very ſlowly.

Mortar made with coarſe Thames ſand, fine Thames ſand, flint-powder and lime; or with fine Thames ſand, fineſt ſand, flint-powder and lime; or with the fineſt ſand, flint-powder and lime; is the worſe, as the quantity of flint-powder is greater, relatively to that of the ſand.

Upon the whole it appeared, that the fineſt ſand is injurious in mortar which is expoſed to the weather, and that flint-powder is ſtill worſe: but that this laſt may be advantageouſly uſed in compoſing ſtucco for inſide work, in which, a fine texture, pleaſing colour, and ſmooth furface, are preferred before extreme hardneſs; and in which the drying may be regulated ſo as to prevent the incruſtation from cracking.

Instead

INSTEAD of resting satisfied with the bare discovery of the fact, that very fine sand, or quartose powder, is incapable of making so good a cement as may be formed with coarser sand, although fine Thames sand and lime make a better cement than can be composed with the coarse sand and lime; and that the mixture of very fine sand, or siliceous powder, with the Thames sand, is rather injurious than useful, although the mixture of the fine with the coarse Thames sand, is better for mortar, than either of them unmixed; I took a great deal of pains to learn the cause of this, in order to confirm or correct the foregoing notions, and render the precepts which flow from this fact, the more satisfactory.

By sorting my finest sand into divers parcels, in sifting it through different sieves; by measuring the meshes of these; and by viewing the grains of each parcel ranked closely on a scale; I perceived, more clearly than I had done before, the roundness of this sand: I moreover found that the grains of the coarsest parcel were, at a medium of their respective bulks, upwards of sixteen times

times larger than thofe of the fineft parcel, the grains of the other parcels being of divers intermediate fizes. As this fand therefore has every advantage attainable by the mixture of coarfe and fine grains, and every difadvantage refulting from the fmallnefs and roundnefs of its grains, I learned the reafon why the defects attending fuch fine round fand in mortar, are not corrected by any mixture of coarfe fand. How thefe defects are induced by the fineft fand and flint-powder, we may conceive in the following manner.

Having already fhewn how the roundnefs of fand tends to render the mortar made with it defective, I may, without any further illuftration of this matter, reckon on this figure of the grains of fineft fand, as one caufe of the imperfection of the mortar in which it is ufed.

There is nothing to prevent the laminæ of lime pafte, which intercede the grains of fineft fand, from being as thick, in the mafs of mortar made with it, as they are in mortar made with coarfer fand; but they are likely to be thicker in general, as the faces of the finer

finer grains are rounder: The number and extent, moreover, of thefe laminæ, muft be greater, in the fum, in the fineft fand than in the coarfer: and it is for thefe reafons, that more lime pafte is required to make mortar with the former than with the latter; fince the mortar is not formed, until the pafte envelopes every grain, and fills the interftices. In this view of the fubject, we difcover another caufe of the defect lately mentioned. If we ufe lime with a fparing hand, it will not extend between all the grains or fill the fpaces; we find the mortar too fhort whilft frefh; and it is as defective in ftrength, when indurated, as it is deficient of the cementing matter. When we ufe the neceffary quantity of lime, the calcareous matter bears a greater proportion to the quartofe grains, in this fineft mortar, than in the coarfer; and this renders the former defective, according to the principles of aggregation already expreffed.

A THIRD caufe of the imperfection of mortar, made with fineft fand, or containing a large quantity of it, appears, on the confideration of the quantity of lime. We have repeatedly feen that mortar contracts the more in drying, and is the more apt to crack, as it contains a greater quantity of lime paste: and as the finest sand requires an extraordinary quantity of the paste to form it into mortar, the aggregation of such a cement is likely to be impaired by fissures, although they do not always appear, by reason of their smallness.

OTHER causes, of the experienced imperfection of fine mortar, might be added, which have no relation to the figure of the grains of sand or the quantity of calcareous matter; but to avoid an excess of theory, I forbear to mention them, and shall only add a conjecture concerning the finer cements.

WHEN a cementious mass, like mortar, is cut with an edged instrument, or broken across, we may observe that the fracture happens in the shortest line, along the laminæ of the weaker cementing matter, and seldom or never in the shorter right line passing thro' the harder grains and the cement alternately, although the impressed force tend to cause the solution of continuity in the shortest, or in a right line. By the principles of mechanics,

the

the refiftance to fuch forces is greater, *cæ-teris paribus*, as the line of fracture is longer, whether it be ftraight, or winding in any courfe: and it it for this reafon, that a wall built with very large fquare ftones, is lefs liable to crack, although the foundation fhould fail near one extremity of it, than a brick wall built with the fame kind of cement on the like ground; or that a wall, whofe bricks are jointed in the prefent fafhion, is more fecure from cracking, than that which fhould be built, on the like infirm ground, with the fame kind of mortar and bricks ftanding over each other, not jointed, but with their fides and ends flufhed, as the workmen exprefs it.

As cements are cut and broken in the direction of the cement, and not in the fhorter line; as the cracks in ill-founded walls run winding along the joints, inftead of going in the fhorteft courfe through the bricks and joints alternately; and as the refiftance of fuch cementious maffes, eftimated by mechanical theory, is greater, as the line of fracture is neceffarily elongated by the ftronger aggregation of certain parts of them; I am inclined to think that calcareous cements made with

lime

lime and quartofe matter, will always be found weaker, under the trial by the chizel or by fracture, as the quartofe fand or powder is finer; becaufe the line of fracture, which takes the courfe of the cementing matter, is fhorter, in any equal depth of fuch maffes, as the hard quartofe grains are finer and rounder.

As flint-powder confifts of exceedingly fine grains of filicious ftone worn to roundnefs in the grinding, what has been faid of the fineft fand is fufficient to fhew, why the cements, which contain flint-powder, are the worft of all thofe we have mentioned.

The cuftomary method of wafhing fand, even for ftucco which is to be expofed to the weather, confifts in paffing it through a fieve, by a circular horizontal motion of it, in a tub filled with water, which flows over, and carries away with it any light matter which can be long fufpended in water, as faft as the fand runs through the fieve into the veffel. But this procefs is inadequate to our views; becaufe the fineft fand fubfides along with the beft, and thefe, in the pre-
cipitation

cipitation, entangle and carry down with them a great deal of finer powder or dirt. Where fuch a method muft be purfued, for want of other utenfils, or through the fcarcity of water, the fand ought to be agitated again in fmall parcels, with a part of the water which has cleared by fubfidence; and immediately after the agitation, the muddy water ought to be poured off, before the light parts have time to fubfide in it. But the ufeful part of the fand is more effectually freed from the finer and noxious parts, by fifting it in ftreaming water, whofe current is to be fo managed, that it fhall carry away the mud and the fand which is too fine, whilft the better part fubfides in a proper receptacle. This art may be gathered from the practice of the Cornifh miners, in wafhing their pounded ores, better than from any written precepts.

IN the fubfequent pages I propofe to fhew the integrant parts of gravel, and their feveral properties in mortar: for my prefent purpofe it will be fufficient to obferve, that the gravel commonly employed in building confifts chiefly, after it is fkreened, of rubble,

coarfe

coarse sand, fine sand, and finest sand, similar to those used in our experiments. This is obvious on the bare inspection of it, and leads us to discover another cause of the weakness of our modern cements, in the composition of which, no other precaution is used, respecting the gravel, except to separate the stones and coarsest rubble from it, by skreening.

When it happens that the skreened gravel contains more than a certain quantity of rubble, relatively to that of coarse and fine sand similar to those described, the mortar made with it must, according to our experiments, be defective. It will be so likewise, whenever the coarse sand of it predominates over the fine sand, to a greater degree than that which was found consistent with the perfection of mortar: and when the quantity of finest sand happens to be considerable in gravel, the mortar made with it must be faulty in a greater degree. Now supposing the gravel to be freed, by the skreening, from every thing more injurious than finest sand and quartose powder, we perceive that the artist, who is ignorant of the advantages of

sizing

fizing his gravel, and ufes it in its native ftate, as chance prefents it, has the odds greatly againft his making good mortar, although he may fometimes do it, without knowing the reafon, as we fhall find hereafter: for his chance is, that the native gravel fhall confift of coarfe and fine fand mixed in the proportion of 4 to 3, or of the rubble, coarfe and fine fand mixed in the proportions above recommended; and that it fhall contain little or no fand like our fineft fand: but the chances againft him are as numerous as there are other diftant proportions of rubble coarfe and fine fand in gravel, and as the kinds of gravel ufed are, which contain the fineft fand or ftill finer quartofe grains, in efficient quantity.

In great cities, where gravel cannot be procured fo cheap as the rubbifh of old walls, which the workmen lay in the ftreets to be ground to powder by the paffing carriages, they ufe this rubbifh fkreened, in the place of fand or gravel, in making mortar. It confifts of the grofs powder of bricks, and of mortar indurated, as much as bad mortar can be, by time; and fome builders affirm that it is

better

better than sand or gravel, for mortar. It is certainly eligible when the price is chiefly confidered; in any other view, it is not fo. From my paft experience I judged the calcareous powder of an old cement, and that of the bricks, to be a brittle perifhable and weak fubftitute for grains of fand; and the quantity of duft in fuch ground rubbifh, to be highly injurious: but as the opinion of the workmen was againft me, I made fome trials of it.

I FOUND that lefs lime was requried to make fat mortar with this ground rubbifh, than with my beft mixtures of fand; which is no fmall recommendation of it in certain jobbs, and is owing, in my opinion,, to the ground calcareous pait, which, fo far as it is finely powdered, is equivalent to whiting: but the mortar made with the rubbifh appeared, in every ftage of induration, and in every comparifon except that of the plafticity, to be greatly inferior to that made with mixed fand and lime, in the fame proportions.

IF the workmen would confine their opinion to the comparifon of fuch rubbifh-mortar

tar, with that in which clayey gravel is used, or with the cements made with the ashes and ordure of the town, dug out in preparing foundations of houses, in those places which were formerly receptacles of such matter, they might maintain it on divers grounds which will be examined hereafter; but otherwise it is erroneous.

SECTION

SECTION XIV.

Experiments made on a larger Scale with our best Mixture of Sands Lime Water and Lime.

IN the spring and summer of the year 1778, I repeated a great number of the foregoing experiments, particularly those which exhibit mortar in the improved state to which I had brought it; and finding my former observations to be true, when the circumstances were not varied, I resolved to try my best cements in larger quantity and in other circumstances. I applied them in the way of stucco, on the brick walls of houses, in different aspects, but chiefly in that of the meridian sun; covering a square yard at least with each specimen, after I had repeatedly wetted the wall with lime water.

By these trials I found that mortar, made with four parts of coarse sand and three of fine wetted with lime water and beaten up with one of my lime flaked with lime water, although

though it could be easily spread on a horizontal plane, or used in building with bricks, was rather too short for plaistering on the perpendicular surface of a wall. It might however be laid on, in small successive portions, by a dexterous management of the trowel, and especially by sliding the instrument on it upwards.

When the weather continued temperate and dry for eight or ten days after the incrustation was made, and no great quantity of rain fell for three or four weeks afterwards, this stucco answered my expectations; for it did not crack in the least, and in three months was almost as hard as Portland stone, at the surface, where the induration first takes place for the reasons formerly mentioned; but it was too coarse to represent a fine grained stone.

Having made two pieces of incrustation of this kind, on the same wall, and knowing that calcareous cements cannot harden so soon as it is necessary in outside stuccoing, unless they be pervious to acidulous gas, in which case they may drink in water likewise, I fre-

frequently wetted one of the pieces, in about three months after it was formed, with lime water, expecting that the calcareous matter of it would cryſtallize in the cement and render it cloſer and harder. I was not diſappointed; for in the courſe of a month I found this piece of ſtucco harder and cloſer than the former, and at the ſurface, as much ſuperior in theſe particulars, to Portland ſtone, as the other was inferior to it. I have ſince found that lime water has not this effect, if the incruſtation be wetted with it, before it is quite dry and indurated ſlowly, to vie with Portland ſtone in that kind of ſtrength which is tried by grinding Portland ſtone on it, or ſcraping it with a chizel; for any other trial of incruſtations is unfair, until the induration has proceeded equally through the whole maſs.

When the incruſtations made of the ſame cement, were wetted by rain, in two or three days, or ſooner after they were applied, and eſpecially when the wind blew the rain forcibly upon them, they were ſenſibly injured, for they never afterwards looked or hardened ſo well as the former ſpecimens of ſtucco.

IN thefe particulars the large incruftations agreed with thofe made on tiles. But the fame agreement did not appear in the incruftations which I had made with the fame compofition, on a wall which fronted the meridian fun, at a time when the weather was very hot; for thefe fhewed a few flender cracks in the courfe of three days. When in the fame fituation and weather, and on a coarfe ftucco of this kind, I fpread, in about two hours after it was laid on, a thinner coat of cement made with finer fand, in order to reprefent a finer grained ftone, the incruftation confifting of thefe two layers, cracked more than the former.

AFTER many repetitions of thefe experiments, in the hotteft weather, with the fame event, I perceived that the trials of fuch cement on tiles, are not fo fevere as thofe to which they may be expofed fometimes in incruftations on walls. In this latter cafe, the ftucco is very unequal in thicknefs; for in the hollow joints and depreffions of the bricks, it is near an inch thick, when over the prominences it has not more than one-eighth of this thicknefs; and as it dries fooneft in the thin parts, the unequal contraction seems to be the cause of those cracks, which would not happen to the same cement laid on the flat surface of a tile: it seems moreover that such a composition may more easily contract in drying, without cracking, as the crust is made narrower or less extensive. But I impute the cracking chiefly to the foregoing unequal contraction accelerated not only by the heat of the sun and the wall, but by the thirsty bricks; for if we form our judgment according to the quicker or slower progress of the exsiccation, and the stiffness which the cement acquires in the act of spreading it on the brick wall, the wetting of this last superficially with lime water, is not equivalent to steeping the tiles for a few minutes in the same liquor.

When, with the view of preventing fissures, I stuccoed a part of the same wall wetted with lime-water with cement containing the mixed sands and lime in the proportion of fifteen to two, in the same kind of weather, I found the difficulty and waste, in applying it, greater than in the former instances, and that it was defective in strength and closeness, for want of lime, although it
did

did not crack. When, through diftruft of my former experiments, I ufed more than one-feventh of lime, the cracks were ftill larger and more numerous.

To guard a recent incruftation from the rain, and to fecure it from cracking in the circumftances laft defcribed, I propofed the expedient of hanging fail cloth on the cornices and fcaffolding: but the expence of this meafure, and the danger of it in windy weather, were ftrong objections.

Embarrassed by this unexpected difficulty, I refolved to change my ground, and try what might be done by a new feries of experiments, in which I intended to ufe every known cheap fubftance, whether it could be reafonably fuppofed to have any confiderable effect towards fecuring a recent incruftation againft the above-mentioned impreffions of rain or hot weather, or could be fufpected of rendering the ftucco defective. I profecuted this enquiry with great alacrity, becaufe I was certain that, although I fhould fail in the attempt towards improvement, I fhould learn how in future to avoid thofe things, which being natively blended in certain kinds of lime ftone fand or water, tend to render the mortar made with them faulty. I had already conceived a notion, which I fhall fubmit to my reader before I conclude, concerning the excellence of fome antient cements; but left I fhould be mifled by it, I proceeded, in all the experiments which I am to relate, on the fuppofition that this excellence is owing to fome matter, accidentally introduced in the materials which the Antients found in the diftricts contiguous to their moft durable cementious works, or defignedly blended with their mortar.

SECTION

SECTION XV.

Experiments shewing the integrant Parts of Gravel, the Choice and Preparation of it; and the Effects of Clay, Fuller's Earth, and Terras, in Mortar.

ON inspecting different kinds of gravel used in London and in divers parts of England, in making mortar, I observed that they all contained some clay; and that this was generally coloured with martial matter. In consideration of the frequency of this matter in mortar, I made it the first subject of my present enquiry.

By the art already described, I sorted three bushels of skreened gravel dug up near Portland Place in Marybone parish, into five parcels; one equivalent to our rubble, another to coarse sand, another to our fine Thames sand, another to our finest sand; and the remainder was set apart as clay or bolar earth. I dried all these, and reduced the lumps of clay to an impalpable powder.

Having treated divers other specimens of gravel in the same manner, I found that gravel, freed from the larger pebbles by skreening, may generally be considered as a native mixture of rubble, sands, and clay; and that when the clay is washed out, the residuary parts of the different kinds of gravel, differ in size, sharpness, colour and hardness; those being the hardest which consist chiefly of quartose matter. Judging of gravel according to the precepts derived from my trials of sand, I rank that dug in Marybone amongst the better kinds of gravel, and used no other in mortar.

After a great number of trials of cements made with my best chalk-lime, lime-water and the gravel, or certain parts of the gravel, and applied on tiles and on a wall, I found that those made with the coarse and fine sand of the gravel, separated from the rest of it, and mixed in their native proportions, were the best; that those made with the rubble coarse sand and fine sand mixed in their original proportions, but containing no other part of the gravel, were the next in hardness and the other desireable qualities;

that

that those containing all the parts of the gravel except the clay, in their native proportions, differed in nothing, that I could discover, from these last, for the finest sand of this gravel was not a fiftieth part of the mass of it; that those containing the rubble sands and clay in the same proportions, and those made with the unwashed gravel, appeared on a close examination to be the worst of all these: and those containing the native unwashed gravel mixed with twice its proper quantity of the clay of such gravel, shewed most clearly that clay is highly injurious, by disposing the mortar to crack in drying, to soften in wet weather, and to moulder when the quantity of clay is one-eighth of that of the sand; but in much smaller quantities, it only prevents the cement from acquiring the hardness peculiar to good mortar, and consequently disposes it to perish in a few years.

With my best mixtures of Thames sands lime water and lime, I blended fine fat tobacco-pipe clay, in different proportions, and exposing these specimens, I perceived that the effect of clay is greater as it is purer and fatter.

ter. The specimens in which the quantity of fat clay was one-seventh or one-eighth of that of the sands, mouldered early in the winter like marle.

These appearances were not altogether unexpected: for in experiments formerly made with a view to the improvement of fire-vessels, I had observed that clay adheres but weakly to any hard bodies, however slowly it is dried on them, and that masses composed of clay and sand in divers proportions, never acquired any considerable hardness by the mere drying and exposure to air: It was therefore not likely that clay should add to the strength of mortar; but as dried clay greedily imbibes water and swells with it, and in drying contracts greatly and cracks, if any thing prevent it from contracting equably; and as marle stones, which consist of clay and calcareous earth, moulder in the weather; it was to be expected that clay would be hurtful.

These experiments point out another cause of the defects of the common mortar, and shew that the gravel or pit-sand to be used in any

any valuable building, ought to be freed from the clay by wafhing, which will be found a very cheap operation, even in cities, if the water which carries off the clay, be directed into a place where it may be depurated by fubfidence, for repeated ufe: they likewife direct us in the examination and choice of thefe, and fhew that the viler kinds may be made equivalent to our beft mixture of Thames fand, or nearly fo, by wafhing and forting, and then rejecting the excefs of rubble or fine fand.

I MUST obferve however that fome kinds of gravel cannot be made fit for mortar by this procefs: for the grains of them, which refemble thofe of rubble and coarfe fand, confift of fmaller grains cemented by clay, which is fo far indurated that it cannot diffufe itfelf in the water fpeedily.

FULLERS EARTH tried in the fame manner was found to operate in the mortar like clay, in every refpect, as I might have prefumed, except that the former was lefs injurious than the clay, when the quantities of them were equal.

TERRAS,

TERRAS, which is a volcanic production consisting chiefly of clay and calx of iron indurated together, when it was ground to an impalpable powder, produced the effects of fuller's earth, in mortar, the more sensibly as it approached nearer to be one-seventh of the quantity of sand. The coarser powder of terras had less effect.

A MORTAR made of terras powder and lime was used in water fences by the Romans, and has been generally employed in such structures ever since their time. It is preferred before any other, for this use, because it sets quickly, and then is impenetrable to water: whence some people hastily conclude that it is the best kind of mortar for any purpose. But by experience I know that mortar made of lime and terras powder, whether coarse or fine, will not grow so hard as mortar made with lime and sand, nor endure the weather so well; but on the contrary is apt to crack and perish quickly in the open air. The efficacy of it in water fences is experienced only where it is always kept wet, and seems to depend on the property which the powder of terras has, in common with other

in-

indurated argillacious bodies, and efpecially the boles, but in a higher degree, of expediting the cryftallization of the calcareous matter, by imbibing the water in which it is diffufed in the mortar, and of fwelling, during this abforption, fo much, as to render the cement impenetrable to any more water: it feems alfo that an acid of the vitriolic kind, which is contained in terras as well as in boles, contributes to the fpeedy fetting of this cement, by reducing a part of the lime to the condition of gypfum.

SECTION XVI.

Experiments shewing the Effects of Plaister Powder, Alum, Vitriolic Acid, of some metallic and earthy Salts, and of Alkalies, in Mortar. Practical Inferences.

IN my best mixtures of coarse and fine Thames sand with one-seventh and with larger quantities of lime, I tried the gypseous powder of which plaister of Paris is made; and found it to be injurious in proportion to the quantity of it. The particular effects of gypsum in mortar, were such as might be expected in consequence of our knowledge of the saline nature of it; gypsum being a compound of calcareous earth and vitriolic acid, which is soluble in water, not so freely as neutral salts, but rather like lime : it disposed the mortar to set faster than it could be applied in stuccoing; it contributed very little to the plasticity of it; and the cement was the more apt to soften in wet weather and to perish in time, as the quantity of plaister powder

der in it was greater. The greatest quantity tried was only one-seventh of that of the sand.

ALUM was found very injurious. The acid of alum formed selenite or gypsum with a part of the lime, and thus operated like gypsum or plaister powder; whilst the earth of the alum induced the imperfections which attend the use of clay. The greatest quantity of alum used was one part in ten of the best mixture of sand and lime; and this specimen mouldered, in nine or ten months, like marle.

VITRIOLIC acid, which formed selenite or gypsum with a part of the lime, produced the effects of a quadruple quantity of plaister powder.

VITRIOLS of lead and of tin, being decomposed by the lime, operated like smaller quantities of vitriolic acid: martial vitriol or copperas had the same effect, and induced an olive colour, which was soon turned to that of rust.

VITRIOL

VITRIOL of zinc or white vitriol, and Epsom salt, did not dispose the mortar to set hastily, nor injure it in any particular discoverable during the application and drying of it; for these vitriols are not easily decomposed by lime: but afterwards I perceived that they impeded the induration of the stucco, and disposed it to suffer by the weather, the more as the quantity of either of them came near to be one-tenth of the quantity of sand.

VITRIOLATED tartar, Glaubers salt, and the salts which are found in most of our waters, such as sea salt, nitre, marine calcareous salt, calcareous nitre, and that composed of magnesia and marine acid, were found like Epsom salt to injure the best mortar: so were caustic mineral alkali; caustic vegetable alkali; and liquor silicum. Caustic volatile alkali, which soon exhales by reason of its volatility, had no sensible effect. I did not try argol or mild alkalies, because they reduce the lime to whiting; neither did I use any acid which forms a very soluble salt with lime, for obvious reasons.

KNOWING

KNOWING that the lime which has been employed by soap-boilers to render their barilla and pot-ash cauftic, contains, even after the repeated elixations, a little alkali and vitriolated tartar blended with the calcareous earth; and that the greater part of this laft is reftored to the condition of chalk, by the acidulous gas imbibed from the alkaline falts; I had, in confequence of the foregoing experiments, fufficient reafon to prefume that this refufe matter of the foap-boilers cannot anfwer the purpofes of lime, or improve our mortar. But as a pretence to the contrary is made by fome artifts, and appears in the Builder's Dictionary, and as the cheapnefs of this article is a temptation towards their extending the ufe of it, I refolved to decide this queftion by direct experiment.

AFTER trying, in my ufual method, fpecimens of mortar made with the refufe of foap-lees and my beft fand, in different proportions; and others made with this fand, lime, and the refufe matter, in various proportions; I found the firft deftitute of the moft ufeful properties of good mortar; and the others were defective, in proportion to the

K quantity

quantity of the refuse matter relatively to that of the lime. Whether this matter improves mortar made with gravel and the common chalk-lime, or encreases the defects of it, is a question not worth our notice.

The experiments lately related shew that lime is the more unfit for building and external incrustations, as it contains more gypsum; and I must now remark that most kinds of lime-stone used in England contain confiderable quantities of this matter which is not much corrected in the burning: But as I have, in the second section, enabled my readers to discover this imperfection, I hope I shall be excused from the invidious office of depreciating or recommending any particular lime-stone or manufactory of lime.

The cautions which our last mentioned experiments suggest with regard to the use of water, are especially necessary in this country, where most of the wells and springs abound with one or more of the abovementioned salts; and it is not to be presumed that the quantity of these contained in water which is used for culinary purposes, cannot

be

be injurious in mortar; for I know that selenite, Epsom salt, the very deliquescent salts compound of magnesia and marine acid and of calcareous earth and the same acid, may, together with a little sea-salt, be natively dissolved in water, to the quantity of half an ounce in a gallon, without affecting the taste of it sensibly. When we consider the quantity of water necessary in slaking the lime making the mortar and wetting the thirsty bricks, and the smallness of those portions of salts, whose injurious effects were discoverable in the course of one year, or in a shorter time, we find sufficient grounds for concluding that such saline waters will be found hurtful in mortar, before many years elapse, particularly where it is exposed to moisture. Indeed this has been already experienced of sea-salt, even in the small quantity of it introduced in mortar, when the sand is taken from the sea shore. The easiest method of discovering the quantity of saline matter in water, consists in evaporating it slowly to dryness and weighing the residue: water which deposites calcareous earth as soon as it is heated, ought to be cleared by subsidence or filtering, before the evaporation is compleated.

When a choice can be made, rain water is to be preferred; river water holds the next place, land water the next, spring water the laſt; and waters noted medicinally or otherwiſe for their ſaline contents, ought not to be uſed at all in mortar; for the ſalts contained in them are thoſe which were tried, the vitriolated tartar excepted.

SECTION

SECTION XVII.

Experiments shewing the Effects of skimmed Milk, Serum of Ox-Blood, Decoction of Lint-seed, Mucilage of Lintseed, Olive Oil, Lintseed Oil, and Resin, in Mortar; and the Effect of painting calcareous Incrustations.

AT the same time and in the same mixtures of the best sand and lime, I tried skimmed milk, serum of ox-blood, decoction of lintseed strained, and thick mucilage of lintseed, in the place of lime water.

THE mortar made with any of these was fatter as the liquor was more glutinous, but was as liable to crack as mortar made with water. In the course of a year it appeared that each of these liquors encourages a vegetation to take place on the surface, which gives it an ugly appearance, and tends to ruin it; and that they all prevent the cement from acquiring the experienced hardness of our best compositions, or indeed from having any competition with them in this particular.

The notion therefore which is entertained by the builders, concerning the use of skimmed milk and blood, is erroneous, unless it be confined to the viler kinds of mortar, which may perhaps be improved by them; because a composition of sand whiting and mucilage, grows harder than that of whiting and sand kneaded with water.

It seems to me that glutinous liquors and good lime act reciprocally on each other, in the time of mixing them, to the destruction of their respective characters, and particularly to the conversion of a part of the quick lime into whiting; and that if any kind of mortar is improved by them, it is then especially, when the workman takes advantage of the fatness induced by them, and using less than his customary quantity of lime, secures his work from cracking.

Olive oil mixed with good mortar, or substituted in the place of a part of the lime water,

ter, rendered the cement defective, as the quantity of oil was greater. The greatest quantity used was half that of the lime.

Lintseed oil used in the same manner, makes the mortar fatter, retards the drying of it, and prevents it from acquiring in any time, so great a degree of hardness as it otherwise would have. It was the more hurtful as the quantity of it was nearer to that of half the lime: in much smaller quantities it was less injurious than olive oil. From my observations on this subject, and on the compositions called oil cements, I have reason to conclude, that no oil ought to be used in a cement which consists chiefly of sand lime and water; nor any water or watery liquor, in a cementious mixture, which is moistened and kneaded with oil chiefly.

As lintseed oil whiting and sand make a cement which hardens to a great degree, in dry situations, and abides the weather a long time before the hardened oil relents, it is not improbable that lintseed oil may meliorate mortar made with bad lime. But good lime and lintseed oil seem to injure each other, in forming a kind of faponaceous compound with the lime water of the mortar.

From the experienced effects of faline, gelatinous, and oleaginous matter, I infer that cow-dung, which I have not tried, would impair good mortar. It makes the common mortar fatter, and in that refpect more convenient for pargeting the interior furface of chimney flues: it feems likewife to prevent the parget made with bad lime, from drying fo quickly and from cracking fo much as it otherwife would do; the fibrous part of the dung being capable of contributing largely to this latter effect. On thefe grounds it may be ufeful in bad mortar thus applied, whether it increafes the hardnefs of it or not; altho' it is likely to impair good mortar.

Powder of refin intimately blended with mortar by grinding it with a part of the lime and lime water, was hurtful according to the quantity of it; the greateft quantity tried being one-fourth of that of the lime.

Before I knew the event of thefe experiments I made an incruftation on a wall fronting

ing the south, but shaded from the sun after mid-day, with a cement composed of seven parts of my mixed sand, one of the best stone lime, and the necessary quantity of lime water. As soon as the incrustation was dry, which happened in four days, I painted one-third of it with lintseed oil prepared for painters use, another third with white lead paint, and the remainder was separated from these by a channel cut between them.

AFTER fourteen months the last-mentioned portion was very hard near the surface, and the induration extended deeply in the mass of it, though not in so great a degree of perfection as that of the surface: The painted portions were also very hard at the surface, but internally much weaker than the other.

FROM my observations of these specimens, and of divers incrustations in this city, which being made of bad calcareous cement, have been painted and sanded, in order to fill the cracks and fence them from the weather; I have had sufficient reason to conclude, that an incrustation, made as good as it may be with lime and sand and lime water, is not bettered by painting

ing it as foon as it dries; that this covering retards the induration of it, by cutting off its communication with the air; that it therefore renders it liable to be irreparably injured in wet weather, wherever the water can get behind the paint; and that if paint or oil ought ever to be applied on fuch ftucco, it ought not to be ufed in lefs than a year after the incruftation is made: I likewife found that the painting and fanding of the common defective incruftations, contributes very little to their duration, although it hardens them at the furface; for it does not effectually prevent them from cracking; and it avails very little to paint the cracked ftucco again; becaufe cracked ftucco is always hollow, as the workmen term it; that is, it parts from the wall in the parts contiguous to the cracks, founds hollow on being ftruck with the knuckle, and falls off in a few years, if it be fo thick and large in extent as to break the adhering portions by its weight.

SECTION

SECTION XVIII.

Experiments shewing the Effect of Sulphur, introduced by different Methods, in Mortar.

IN my first trials of sulphur, it seemed to be useful; and this led me to try it in so many different ways, and in so many mixtures of lime and sands, and of these with flint powder and divers other substances, as would render the recital of all my observations on the effects of it inconsistent with the plan of this essay: I must therefore content myself with communicating those which I think most useful, in such terms as may give some intimation of the manner in which the experiments were made.

When the livigated powder of sulphur was mixed with mortar already made of good materials and did not exceed one-thirty-second part of the mass, it seemed to improve it, in the first and second month, and sometimes during a longer time of comparison with mortar made of all the same materials, except sulphur, in similar proportions: But in ten or twelve months the sulphur was found injurious, and the more so, as it exceeded the foregoing proportion. The most hurtful effect of it was, its disposing the mortar to relent in long continued rains, and grow quite friable after a few alternations of freezing and thawing. It had the same effect in mortar containing several of the ingredients already named and of those hereafter to be mentioned.

When the sulphur was mixed with fresh powdered lime, and these were ground briskly with lime water, a calcareous liver of sulphur was formed, proportionate to the quantity of sulphur used; and the mortar made with this mixture and sand, or with this and sand and other ingredients, was worse than mortar containing an equal quantity of the sulphur mixed in it in the former method.

The transparent liquor called liquid calcareous liver of sulphur, which consists of sulphur dissolved in water by the intervention of lime, being used instead of water in making mortar with sand and lime in any proportions, was found more injurious

jurious than three times this quantity of undiffolved fulphur was, in the firft-mentioned method of ufing it: and this liquor had the like effect in mixtures of mortar with divers other ingredients. Whence I infer that fulphureous mineral waters ought not to be ufed in mortar.

IF the plan of thefe experiments had not comprehended the noxious as well as the ufeful ingredients, and I had not refolved to diftruft every theory, I might have prognofticated the event of thefe mixtures, in confequence of my certain knowledge of the operation of fulphur lime and air on each other.

WHEN fulphur and lime are moiftened with water, and expofed to air, the acid of fulphur being attracted by the lime, whilft the phlogifton of the fulphur is attracted by the air, a decompofition of the fulphur takes place, and new compounds are formed. The acid and lime gradually form felenite or gypfum, whilft the air combined with the phlogifton is wafted away. Therefore lime, by fo much of it as is thus expended in forming

gypfum,

gypsum, is not only unable to act as a durable cement of the grains of sand, but is capable, according to the experiments of the sixteenth section, of counteracting the cementing powers of the residuary part of it, when the mass of sulphurated cement is exposed to the weather.

The pleasing warm colour which sulphur induces in calcareous stucco, whilst it is fresh, and the promising appearances of such an incrustation in the first year, have, if I am rightly informed, already misled an artist to apply it freely at his own risque. I wish these observations may serve to undeceive him.

About this time, the imitation of coloured stones, by incrustations, became an object of my attention; and some of the subsequent experiments were made with a view to it, as well as to the purposes already expressed.

SECTION

SECTION XIX.

Experiments shewing the Effects of Crude Antimony, Regulus of Antimony, Lead Matt, Potter's Ore, White Lead, Arsenic, Orpiment, Martial Pyrites and flaked Mundic, in Mortar.

CRUDE antimony reduced to an impalpable powder and then ground with the lime and lime water, operated in mortar as sulphur does when it is used in the same manner and in the quantity which the crude antimony contains. The antimonial powder moreover induced a disagreeable blueish colour, which in a little time became brown and afterwards yellowish.

WHEN the powder of antimony was mixed in the mortar after it was made it was less injurious.

REGULUS of antimony tried in the same way, seemed to have no other effect than that

that which is produced by the admixture of flint powder or other fine powders of hard bodies.

Powdered lead matt and potter's ore of lead acted like crude antimony, but more flowly and weakly in equal quantities of them.

White lead was found exceedingly injurious, which I expected; for I had long before difcovered and fhewn in my public Courfes of Chymiftry, that a great part of white lead is acidulous gas, into which vinegar is eafily convertible in the procefs for making white lead and in many others; and I forefaw that the lime, attracting this matter, would be reduced to the condition of whiting in the time of making the mixture, and that the mortar would confequently be defective. The white lead, as faft as its acidulous gas is drawn from it by the lime, becomes yellow like mafficot. As white lead improves the oil cements, thefe experiments fhew that there is no true analogy between the calcareous water cements and thofe which are called oil cements.

Arsenic

Arsenic operated in mortar like the neutral falts; and orpiment produced the injurious effects experienced of fulphur and of arfenic; which effects were greateft when the orpiment was ground with the unflaked lime and lime water. Orpiment imparted a dark brown colour at firft, which foon became yellow and afterwards difappeared.

The martial pyrites called mundic, heated to rednefs, and then flaked by moiftening it with water whilft it was hot, operated like crude antimony, with this difference only, that a greater quantity of it was required to produce the fame effect; for this reafon, as I conceive it, that the quantity of fulphur in martial pyrites is lefs than in crude antimony, and being held in it by a more forcible attraction, is prevented from acting fo freely on the lime of the mortar. The colour induced by the flaked mundic was at firft blueifh and afterwards turned to that of iron ruft.

The mundic, which was that of Wiel-Virgin in Cornwall, ufed in its native ftate, in mortar, kept me in fufpence upwards of twelve months. It was tried not only on tiles,

tiles, but in large incruftations on walls, becaufe it promifed great advantages at firft. When the quantity of it did not exceed one twenty-fourth of that of the mortar, it manifeftly increafed the induration of the cement during the firft nine months; but after fourteen or fifteen months it difpofed the incruftation to relent, the more as it was oftener wetted or as the place was damp, and from being exceedingly hard, to become penetrable to a pointed inftrument pufhed only with the hand, and as brittle as chalkftone. The colour and changes of colour of the mortar containing native mundic, are fimilar to thofe produced by the flaked mundic, and are not at all pleafing to the eye. The effects of much fmaller quantities of this matter in mortar do not yet appear fo clearly; but there is no reafon to prefume that they will be of the fame kind, though in a fmaller degree.

These and the preceding experiments indicate that all bodies foluble in water, not excepting arfenic, and all thofe which are capable of efflorefcing, or of being decompofed by air and moifture, are hurtful in mortar; and they teach us to avoid thofe kinds of gravel

which

which are impregnated with pyritous matter, whether it be arſenical, metallic, aluminous, or calcareous. The effects of regulus of antimony, and the ſpeedy decay of the cheaper metals, however perfectly they are deſulphurated, give ſtrong grounds for preſuming that calcareous cements, which are to be expoſed fully to the weather, are more likely to be injured than improved by metallic matter introduced in any form.

SECTION XX.

Experiments shewing the Effects of Iron Scales, washed Colcothar, native Red Ochres, Yellow Ochres, Umber, Powder of coloured Fluor, coloured Mica, Smalt, and other coloured Bodies, in Mortar. Advices concerning coloured Incrustations, Inside-Stucco, and damp Walls.

IRON scales from a smith's forge, which consist of iron semi-calcined, and are thought by many artists to improve mortar, were tried eighteen months ago, by grinding them to a fine powder, and mixing this in mortar, to half the quantity of the lime, and in smaller proportions.

THE larger quantities, in the course of twelve or fourteen months, appeared to be hurtful; and by these I judge of the smallest, which do not yet appear to produce any remarkable effect in incrustations made in dry situations, except the rusty colour which they induce. But in those which reached near the ground, and in others made on tiles which were laid flat on the ground in a fhaded damp corner, in both of which inftances the incruftations were always moift, the iron powder feemed to render the cement a little harder than it could otherwife become in the fame time in fuch circumftances, and it certainly made it clofer in the grain.

By thefe experiments I am inclined to think that iron powder, which, during its converfion to ruft, imbibes a great deal of acidulous gas and air, and fwells confiderably, may be ufed with fuccefs, where the proper induration of good mortar is prevented by continual moifture, and the chief purpofe of the cement is, to exclude water perfectly; by the clofenefs of its texture, to which the fwelling of the iron contributes not a little. If it is capable of producing any defireable effects in cements otherwife circumftanced, thefe are to be expected only when the quantity of it does not exceed one-eighth of that of the lime, or one-fiftieth of that of the mafs of mortar.

WASHED colcothar of iron, native red ochres, yellow ochres, and umber, had the effects of smaller quantities of terras, or of equal quantities of flint-powder.

COLOURED fluor and micaceous stones, coloured marble, smalt, and divers other coloured substances, which are insoluble in water, reduced to fine powder, induced their respective tints in the incrustations, but acted like flint powder.

FROM the experienced effects of coloured calces of iron, and of divers sulphurated and perishable metallic powders, I learned that these ought not to be used in external incrustations; since they render them more defective as they colour them deeply; and I turned my thoughts to the discovery of some other expedient for inducing permanent colour without injuring the cement.

I SOON found that this may be done, with regard to the lighter and pleasanter tints, by the use of coloured sands, or the coarse gritty sorted powder of hard and durable coloured bodies. Lynn sand affords a white cement
which

which is the better, as more of the fineſt part is ſifted out of the ſand. Thames ſand makes a grey cement not unlike Portland ſtone, and this colour is agreeably varied by the uſe of grey bone-aſh, of which we ſhall preſently treat.

A RICH yellow tint is obtained by uſing the golden yellow ſands, of which kind there is one near Croydon in Surrey; and a ſmall quantity of this ſand mixed with Lynn ſand, gives a warm white, and with Thames ſand, an exact reſemblance of the Bath ſtone. Theſe are the moſt eligible tints for the fronts of houſes.

UNTIL I had tried the gliſtening ſcaly talcs, I imagined they would ſerve to impart all other tints, as they may be had of any colour, and are as durable as they are pleaſing to the eye: but they were found to weaken the adheſion of the cement to the wall, and to make it ſo rough and ſhort, that it was almoſt impoſſible to form a ſmooth compact incruſtation with it, unleſs the lime were uſed in exceſſive quantity; and in the courſe of eight or nine months it appeared that the cements, in which they were mixed in the quantity neceffary to produce ftrong tints, were rendered fpongy, and greatly weakened by them.

Scaly gliftening mica, ftrewed equably on an incruftation previoufly wetted with a thin mixture of lime water and lime, and gently compreffed to lay the fcales flat, imparts its colour with the fulleft effect. In this way coloured mica may be ufed, where it is cheap, on external incruftations, if the perfpective appearance of a building can be improved by different colours of any members of it: and this kind of colouring greatly excels painting, in the fickle weather of our climate, becaufe it lafts unfaded, as long as the micaceous cruft.

To tinge a cement fufficiently for profpect or contraft, of any colour which is not found in fand, fo that the incruftation fhall not be impaired, and that the colour fhall be as durable as the cement; I found nothing more advifeable than to ufe, in the place of the fand, or of a part of it, coloured glaffes or coloured ftones of the hardeft kind, beaten to coarfe coarse powder, the finer parts of which are to be washed away, not merely because they are injurious to the cement, but because I have observed that they contribute very little to the intended colour.

THE drying, induration, and texture of incrustations made on brick walls and other irregular surfaces, are always so far unequal as to exhibit visible traces, which deform the work and cannot be effectually obliterated by any known method so convenient as that of covering the first coarse incrustation, after it has dried, with another coat which may be made finer and smoother. Thus the expence of fine grained smooth or coloured stucco is rendered moderate; because the finer, or the colouring materials, may be reserved for the exterior coat, which will last for ages, if the cement be good; as we shall shew, when we come to consider the experienced duration of the best calcareous cements.

As the mouldings and paintings which are expended on the soft stucco now used, and which contribute so much to the magnificence of our apartments, can be equalled, in
their

their ornamental effects, by the double incruftations which I have defcribed, and greatly exceeded by thefe laft in the hardnefs and duration of them, I do not doubt that plaifterers will adopt this improved method, when they find that it is confiftent with their own intereft, as well as with that of their employers.

I AM not fufficiently acquainted with their bufinefs to form a juft eftimate of this fubject; but I will fubmit to their confideration a few obfervations which would influence me very much in the choice of ftucco for a houfe of mine.

THE compofitions heretofore ufed for ftuccoing within doors, are incapable of hardening confiderably, and when they are laid on the naked walls, foon become tarnifhed, unfightly, and inconvenient, by the damps which the workmen call fweating, and which are, in my opinion, of two kinds; one I will call damp by tranfpiration, the other damp by condenfation. The damp by tranfpiration occurs, when the principal walls are ftuccoed before they have dried, or when

the

the materials of them are so spongy as to imbibe the rain, and the circulation of air within the house is not sufficient to waft away the moisture which transudes from the wet wall into the stucco; and especially when the exhalation of this moisture from the stucco, is impeded by the closeness of its texture; for all such bodies retain moisture the more forcibly, as their pores are smaller, and as the air meets more difficulty in pervading them. I see no reason to doubt that this inconvenience would be obviated by making the incrustation of a texture similar to that of the materials on which it is laid; and that the cement made with about seven parts of sand, one of lime, and the lime water, and improved, as we shall teach hereafter, by the admixture of bone-ash, would continue dry in such circumstances, because moisture quickly exhales from it, by reason of its texture.

The damp which seizes incrustations, when the walls are badly constructed, when the joints of the facing bricks become hollow by the decay of the mortar, or when the copings

pings or gutters are defective, do not fall under our confideration.

The damp by condenfation appears moſt on the fineſt and cloſeſt incruſtations, however perfect and old the walls may be. To find the proximate cauſe of it, we need only to advert to that which gathers on glaſs windows, whilſt the wainſcoat and other ſpongy bodies, which ſerve to incloſe the ſame rooms, remain dry; or to the moiſture which gathers on walls faced with the cloſer kinds of ornamental marble, in ſumptuous buildings, at the ſame time when the walls and incruſtations, which are contiguous to them, and are of a coarſe texture, are quite dry. In theſe and other inſtances we may perceive, that the damp is owing to the cloſeneſs of theſe bodies, and that a ſtucco pervious in a certain degree to air and moiſture, will be free from it, as well as from the other lately mentioned.

The plaiſterers, finding their ſtucco, which is as fine and cloſe as they can make it, liable to contract theſe damps, eſpecially on the principal walls of houſes, caſe them with lath-work, on which the incruſtation is laid diſtant from the wall. In this way they

they obviate the appearance of damp; but they at the fame time contract the rooms, and narrow paffages and ftaircafes fenfibly, at a great expence. This is enhanced by the repeated plaiftering neceffary to fill the flender cracks which disfigure their incruftation during the drying, and by the oiling or painting which is finally required to hide this defect compleatly, if not to give colour. Thus the work becomes coftly, although the plaifterers profit is moderate.

On thefe confiderations I am inclined to the opinion that it will be found as advantageous to the plaifterer, as to his employer, to prefer our cement before any other, for internal incruftations; efpecially when no other colour is required, befides thofe which may be imparted by coloured fand, or materials which do not greatly exceed it in price. I would not interfere with the workman, in forming an exact comparative eftimate of the expences, if I could do it; but I will venture to affirm that an incruftation made as I have defcribed, or in the improved method hereafter to be fhewn, will be found ultimately cheaper than any other yet difcovered, for
the

the following reasons. It will be more durable by reason of its greater hardness; it will retain its colour longer unfaded, because the colouring materials do not tarnish or perish like paint; it will preserve the sharpness of the mouldings and the elegance of its appearance longer, because it will not require the frequent painting which soon blunts the figures and mouldings of ordinary stucco; it may be finished with less labour, because it is not apt to crack in these circumstances, and does not need many coats and repeated plaistering; and as it is not likely to contract damp, it will save all the expences and inconveniences of lath-work, whether it be laid on partitions or on principal walls, provided the cement applied on the former be not made of the finest materials.

If a polished and white surface of our stucco should be required, it ought to consist of two layers: the first of which is to be coarse and capable of hardening to the highest degree; the second is to consist of flint powder lime and lime water, and is to laid on very thin, and finely smoothed. To give a rich colour together with a smooth surface, to our

best

best incrustation, we must use, in the place of flint powder, for the finishing coat, the coloured powder of sands, or stones, or glasses; and introduce as much of the colouring ingredients used in painting, as will be sufficient to give the required appearance, avoiding those which are spoiled by lime.

To my eye, the warm white, or coloured stucco which is not quite smooth, is the pleasantest: but those who prefer the smoothest, may have it made at a moderate expence, in this last-mentioned method, in which the useful and solid part of it, contributes to the support and duration of the weaker ornamental coat, which thus circumstanced is likely to preserve its beauty for a very long time, although it might, in the weather, be impaie d in three or four years.

SECTION

SECTION XXI.

Experiments shewing the Effects of common Wood-ashes, calcined or purer Wood-ashes, elixated Ashes, Charcoal Powder, Sea Coal-ashes, and powdered Coak, in Mortar; and Observations on their integrant Parts, and the Differences between them and the Powders of other Bodies.

THE ashes of wood and sea-coal are frequently mixed with mortar, or used in the place of sand, in laying tiled floors, and even in external incrustations. Some workmen say they are used in the former case to save sand; others that they serve to resist moisture; and those who seem to be the best informed affirm, that they hasten the drying and induration, and prevent the cracking of mortar which is laid very thick in order to fill the depressions of walls which are to be stuccoed; and that they are used in finer incrustations with the sole view of preventing cracks.

THE

The ashes of the same kind of wood differ, according to the circumstances in which they are formed, even upon the same hearth, not only in colour, but in other particulars known to chemists, which I shall attend to presently. As the separation of these different sorts of ashes is not practicable at a moderate expence, and never is attempted by the workmen, I contented myself, at first, with procuring the ashes of cleft pollards burned on a hearth, and with sifting the whole quantity of them, to free the finer part from the fragments and coarse powder of charred wood which formed a great part of the bulk of them. The sifted ashes were grey inclining to brown, strongly alkaline to the taste, and viewed through a convex lens, were found to contain a considerable quantity of fine charcoal-powder, which I estimated at one sixth or more of their bulk.

To learn the effect of the purer ashes, or of the more dephlogisticated earthy and saline parts separated from the charcoal, I took about a gallon of the sifted ashes, and burned them on a test in a reverberatory furnace, with a heat not exceeding that of a culinary fire

fire, taking care to accelerate the combuftion of the charcoal powder contained in them and render it equable thro' the whole heap, by ftirring it, and prefenting frefh furfaces to the air, until the whole was rendered incombuftible. After this procefs, the powder, which I fhall call calcined wood afhes, was rather brown than grey, and retained its faline tafte.

On trying the fifted wood afhes in my beft mortar, and in other mixtures of fand and lime, I found that they gave the cement a fpongy texture, and enabled it to dry without cracking, when the lime was not ufed in exceffive quantity; but that they prevented it from acquiring the hardnefs of mortar made of lime and fand only: fo that the advantages which they promifed to afford in certain circumftances, appeared to be counterbalanced by the permanent weaknefs induced by them; which latter effect was the greater as the quantity of the afhes came nearer to equal that of the lime.

The calcined wood afhes likewife prevented the mortar from cracking, without making it fo fpongy: but they manifeftly impeded

peded the induration of it, and difpofed it to be injured by rain, in the fame manner as fmall quantities of alkali were found to do.

On a ſtrict comparifon, the calcined wood afhes, which we may confider as afhes freed from charcoal powder, appeared to be much more injurious than the uncalcined. This I imputed to the greater quantity of alkali in the former, which is hurtful in a double capacity; firſt as a faline body; and fecondly as a compound which yields its acidulous gas to lime, in the inftant of mixture, and confequently impairs the cement.

Mortar made with bad lime in the ufual proportions may neverthelefs be improved by fifted wood afhes; for the coal and earthy part of thefe, if they were only equivalent to fo much fand, render it lefs liable to crack; and the bad effects of the alkali may be greatly overbalanced by this advantage, in an incruſtation which is required to be rather uniform and fecure from cracking, than hard and durable in the higheſt degree.

I MUST not omit this opportunity of observing that calcined wood aſhes, and even the ſifted freſh wood aſhes, improve plaiſter of Paris in hardneſs, to a very great degree, if it be kept in a dry place. The ſolution of this phænomenon is not difficult.

ANY perſon who intends to repeat my experiments on calcined wood-aſhes, ought to take care that they be not calcined with a ſtronger heat than I deſcribed; for if he exceeds this, the aſhes, after the ſigns of their combuſtion have ceaſed, will ſmoak ſtrongly, a part of the ſaline matter being ſublimed in the mean time; and the remaining earthy and ſaline portion will form a light grey or brown ſemivitrified gritty powder, or will concrete into lumps. This matter will then be found inſipid and equivalent to ſand, in mortar, as I have experienced; for it differs as much from wood aſhes, as the powder of potters-ſtone-ware differs from the raw clay.

WHILST I was employed in theſe experiments, the following thoughts occurred to me. The aſhes uſed by the workmen, being paſſed through a coarſe ſieve, may conſiſt

for

for the greater part, of charcoal, which afterwards is beaten finer in making the mortar: The ashes used by builders whose durable works authorized this practice, might have been the refuse of manufactories of potash, in which the saline matter is always carefully extracted from them; and charcoal powder or elixated ashes may greatly improve mortar, altho' ashes finely sifted and replete with salts should impair it. I therefore boiled my calcined wood-ashes in water, and repeated this operation twice in fresh water; knowing that one elixation does not free the ashes perfectly from the saline matter: I then dried the insipid ashes thoroughly, and used them in this state, under the name of elixated wood-ashes. At the same time I provided charcoal powder sifted thro' the same sieve which I used for the wood-ashes.

After a great number of experiments made in the usual manner with the elixated ashes, I found that they rendered the mortar spongy, disposed it to dry and harden quickly, and prevented it from cracking, more effectually than the like additional quantity of sand would do it. They did not appear to induce the defects attending saline bodies in mortar; they only made it weaker, as the quantity of the elixated ashes was greater relatively to that of the sand or lime. This weakness, however, was not such as the unwashed ashes or saline bodies produce, but rather of the kind which I pointed out in those parts of the foregoing sections, wherein I endeavoured to shew, that cementious masses resist edged instruments or any force tending to break them, the more weakly, as they contain more of the softer and brittler calcareous matter, or as softer grains are substituted for a part of the sand.

In every comparison of the specimens containing unwashed wood-ashes, with those in which the elixated ashes were mixed in the same proportions, it clearly appeared that the latter are to be preferred; and that neither of them ought ever to exceed half the quantity of lime, in good mortar.

As flint powder and other earthy powders were found to dispose mortar to crack, I could not conceive how the elixated wood-ashes operated so effectually in preventing

this

this defect, until I examined them attentively, and found them to differ from the other powders in two particulars. Elixated wood-ashes contain very little powder of the finer kind; they feel gritty between the fingers, and appear to consist of ragged spongy small grains compressible to a considerable degree in the heap. How a powder thus conditioned prevents the cracking of mortar or otherwise improves it, I shall attempt to explain, after stating other facts upon which my notions of this subject are founded.

CHARCOAL powder had the same effects as elixated wood-ashes, with these differences only, that the cements containing the larger quantities of charcoal powder could be more easily cut, and were of a bluer colour, than those containing the like quantities of elixated wood ashes. The powder which I used was sifted like the ashes; and, viewed through a microscope, answered to the description lately given of elixated ashes,

THE skreened ashes of Newcastle coal consist chiefly of charred coal or coak, and as they contain very little saline matter, are insipid.

When I reduced them to powder, and paſſed them thro' the ſieve, they anſwered to the deſcription given of elixated wood aſhes, and produced nearly the ſame effects in mortar. They did not weaken it ſo much as charcoal powder had done; which I impute to the greater hardneſs of the ſmall grains of coak.

In all theſe compariſons, it is to be underſtood that I made them at the ſame periods of the induration of the ſeveral ſpecimens.

From theſe experiments I conclude that, where a choice can be made, theſe powders are eligible in this order; elixated wood aſhes freed from the fineſt powder in waſhing, firſt; powdered coak or ſea coal cinders, next; charcoal powder next; rough wood aſhes powdered, laſt: But well burned fine unwaſhed wood-aſhes ought not to be uſed at all in external cementious work or incruſtation.

The laſt of theſe gives a diſagreeable grey or duſky colour to the cement; and the others, a blueiſh or ſlate colour, ſtill more offenſive to the eye; for which reaſon they

are

are unfit for any work that is not hid from the view.

As my reader may not fully underſtand what I briefly mentioned concerning the ſenſible difference between theſe laſt examined powders, and others noticed in the preceding ſections, I will thus exemplify my notions. Wood conſiſts of watery and volatile parts which are expelled by heat, and of fixed parts which conſtitute the charcoal: and charred wood, which greedily imbibes air or water in great quantity, may be conſidered as an aſſemblage of capillary tubes of divers figures and ſizes. So we may likewiſe conſider the fragments of charcoal, and each viſible grain of its powder. But as the moſt brittle bodies are flexible when they are made ſufficiently thin, the charcoal powder is an aſſemblage of ſmall flexible or compreſſible tubulated bodies.

As the charcoal which is the more fixed and ſolid baſis of wood, is ſpongy after the juices are expelled in charring; ſo the aſhes of charred wood are, after the elixation, an aſſemblage of ſpongy or tubulated grains out of of which the phlogiſtic matter has eſcaped during the combuſtion: and the texture of theſe grains differs from that of the grains of fine ſand or of flint powder, in the ſame manner, if not in the ſame degree, as the texture of ſponge differs from that of a flint. And we may conceive the unwaſhed wood aſhes, as a heap of ſmall ſpongy bodies clogged with alkaline ſalt.

Upon the ſame grounds, the relation of coak or ſea-coal-cinders to the raw coal, is analogous to that which charcoal bears to wood, or ſpongy pumice ſtone to porphyry; and transferring theſe obſervations to bones, and conſidering the ſmaller veſſels and finer texture of them than of wood, we ſhall find the powder of charred bones to conſiſt of tubulated or ſpongy bodies like thoſe of charcoal powder, but pervious by ſlenderer and harder tubes; and bone-aſh, which is the gritty powder of well burnt bones, to have the ſame relation to the charred bones, which elixated wood-aſhes have to charcoal powder.

Thus I have thought of theſe ſubſtances, after having obſerved what happens to them

in

in the preparation; examined them by a microscope; experienced their effects to be so different from those of finest sand, or powdered stones, in mortar; and finally discovered, by repeated experiments, the detail of which is not now necessary, that semivitrification, which destroys the spongy texture, and levigation, which breaks these spongy grains down to the particles of which they are constructed, render charcoal powder, wood-ashes, powdered sea-coal-cinders, and others of the like kind, incapable of acting in the manner described, in calcareous cements.

All these things being considered, I impute the effects of these ashes, or powders, to the tubulated structure and compressibility of the integrant parts of them; and in the next section I shall offer all that I have attempted further, theoretically or practically, relative to this subject.

SECTION XXII.

Experiments shewing the Effects of white and grey Bone-ashes, and the Powder of Charred Bones; and Theory of the Agency of these in the best calcareous Cements.

LONG before I had tried all the powders heretofore mentioned, I used bone-ash in many experiments, and saw the effects of it in mortar. For the sake of brevity and perspicuity I reserved the relation of them for this section: and in order to shew more clearly the analogy in texture, between bone-ashes and the powders lately mentioned, and to suggest the means of procuring them in any part of this country, I will premise a sketch of the most profitable processes by which they are prepared, at a moderate price not much exceeding that of good stone-lime.

The bones collected in great cities, are broken to small fragments in a mill, and boiled in water, in order to extricate and save the oil of them. They are then put into a large iron still, through an aperture which is stopped up closely after the charge is made. The still, which opens into an apparatus of refrigeratory vessels, is heated gradually to redness, until all the volatile alkali, commonly called spirit and salt of hartshorn, is expelled from them, together with empyreumatic oil, water, and certain elastic invisible fluids: The alkali, being the only valuable article amongst these, is retained and condensed in the refrigeratory tubes and vessels with all possible care, whilst the elastic fluids, left they should burst the vessels, are suffered to escape in places distant from the fire or the flame of candles, because they are combustible, and if they catch fire whilst air remains in the condensing vessels, explode like gunpowder.

The bones thus heated without being exposed to the air, are charred to blackness, but still remain combustible. When they are required in this state, the iron still is
kept

kept clofed until they cool, and then the blackeft of them are ground to fine powder, which is ufed as a fubftitute for ivory black, which is prepared in the fame way from ivory. The coarfer powder of thefe, is what I underftand by powder of charred bones. But when this is not the manufacturer's defign, the door of the iron ftill is opened whilft it is hot, and the charred bones, which flame and burn when they meet the air, are thrown into a kind of kiln, at the bottom of which the air can freely enter, and maintain the combuftion, until the bones are burned to whitenefs, for the greater part. The white fragments are picked, and rather bruifed, than ground, to a gritty powder, by a millftone which rolls on them vertically over an inclined circular plane. This powder paffed thro' a fieve is called bone-afhes, which are much ufed in metallurgy, and fitter for our purpofes in incruftations, than the powder of burned bones ground as pigments are. The fragments which have not been thoroughly burned in the kiln, form a dark grey powder; and mixtures of the white and grey burned bones afford bone-afhes of the lighter grey colours.

THE

THE whole quantity of bone-afhes, which is to be ufed in the fame incruftation, ought to be well mixed ; for it is impoffible to fort the well burned or the grey bones fo accurately as to fecure an unity of colour in the parcels of powder which are fucceffively prepared, and a very fmall variation of colour will be feen in the incruftation.

MR. JOHN OLIVER of Hoxton, who is a very liberal and ingenious artift, prepares bone-afhes, judicioufly adapted to the purpofes which I am now to mention, at a very low price, as well as the coarfer kind which is ufed in making cupels and tefts by the refiners. I fhall diftinguifh the former by the name of forted bone-afhes ; becaufe they are freed from the fineft and the coarfeft parts of the latter.

As I knew that bone-afhes confift chiefly of calcareous earth, and may be reduced to lime, by diffolving them in acids, precipitating the folution by alkalies, wafhing the precipitate perfectly and then burning it, I tried them with fand in different ways, in order to learn how far they refemble lime in their ce-

menting

menting properties; and found that the sorted bone-ashes had very little effect; but that compositions made with the levigated powder of these and sand and water, were nearly equal in hardness to those made with whiting and sand kneaded with water in the same proportions, and were not so liable to crack. Hence I inferred that bone-ashes, of which five-sixths are calcareous earth, could not improve mortar by any augmentation of the cementing powers of the lime, although they might be useful in other respects, and that they could not supply the defect of lime in quantity or quality.

In the course of two years I made so many experiments with bone-ashes mixed in mortar composed of lime sand and lime water, in different proportions, and of these with divers other ingredients, that I may venture to say I attained a thorough knowledge of their effects, and need not hesitate to relate them in the style of precept.

The sorted bone-ashes, mixed with mortar in any quantity not exceeding that of the lime, dispose the cement to set speedily with-

out

out cracking; and effectually fecure it from cracking, if it does not contain lime in fuperfluous quantity: They likewife give a texture which is the more fpongy as the quantity of the bone-afhes is greater; and they accelerate the induration of it through the whole mafs.

THE forted bone-afhes encreafe the plafticity of frefh mortar which is made with the fmaller quantities of lime in order to fecure the work from fiffures; and thus they are ufeful in a triple view, in external incruftations; by facilitating the operation of plaiftering, by preventing cracks, and by bringing the incruftation quickly to a ftate in which it is not eafily injured by unexpected rain.

WHEN the forted bone-afhes exceed the lime in quantity, they fenfibly injure the cement, by rendering it weaker. How thefe afhes, which are not equivalent to fand in the hardnefs of their grains, nor to lime in their cementing powers, operate to weaken the cement, may eafily be conceived, in confequence sequence of the observations made in the ninth, twelfth and thirteenth sections.

When the sorted bone-ashes are mixed in mortar in the quantity of one-fourth of the lime, they improve the plasticity, if the mortar be short, and they produce the desireable effects above mentioned in a sensible degree, without weakening the cement in the same proportion. As a smaller quantity of them seems to be useless, and a greater quantity than that of the lime injurious, the following rules are to be observed.

When the artist is more solicitous to secure an incrustation from the effects of hot weather, to finish it quickly, to hide the traces of brick-work which are apt to appear thro' it, and to guard it against rain, than to make it hard and durable in the highest degree; he is to use as much of sorted bone-ashes as of lime: When the season, exposure, and other circumstances permit him to attend solely to the true excellence and duration of his work, he is to use, in our best calcareous cement, only one part of the sorted bone-ashes for every four parts of lime. By these
rules

rules he may chuse intermediate quantities adapted to his purposes.

The coarser bone-ashes used in making cupels and tests, do not go so far, so the workmen express themselves, or do not operate so effectually, as the sorted ashes, in equal quantities of them by weight; and finer or levigated bone-ashes are rather injurious than useful in the coarser cements.

The black powder of charred bones, and grey bone-ashes have nearly the same effects as sorted bone-ashes have, when the powder of them is sorted in the same manner; excepting what relates to colour.

These observations on bone-ashes were made before the expiration of 1777, on specimens of mortar laid on tiles, and small pieces of incrustation made on the walls of my house and on the fence-walls behind it: But they were not thoroughly confirmed until a comparison was made between large incrustations laid in trying aspects and containing bone-ashes, with those made close by them of my best mortar, in the year 1778;

when I discovered the difficulty expressed in the fourteenth section, of making extensive incrustations, in certain circumstances, so free from defects, as the smaller ones were which I had made at home.

In this last mentioned year I was favoured by Mr. James Wyatt of Queen-Ann-street Cavendish-square, the celebrated architect of the Pantheon, and by his brother Samuel Wyatt of Berwick-street, who is a very eminent builder, with the best opportunities of making these comparisons: for they ordered the plaisterers employed under them to apply my compositions in a workman-like manner, in different aspects and in large quantity, and thus enabled me to judge truly of the merits of them.

By the analogy of bone-ashes to cinders or ashes of other bodies, by the effects of them in my experiments, and by the observations which I have made on capital houses and garden walls which have been fronted or entirely stuccoed with my cement, some in the months of October November and December in the year 1778, others in the
Spring,

Spring, the hotteſt weather, and the Autumn of 1779; I have been led into the following opinions concerning the agency of bone-aſhes in calcareous cements.

THE mortar which contains bone-aſhes, partakes in ſome degree of the compreſſibility and ſponginefs of their grains, and is the leſs liable to crack in ſetting, for the ſame reaſon that ſponginefs is, in any other body, an effectual preventative of fiſſures in drying; or becauſe, any contraction of the lime paſte, in conſequence of the exhalation of its water, is confined to the circuit of the ſpongy grains compreſſed in beating trowelling and floating the cement, and is thereby prevented from running longitudinally to form fiſſures. The ſame texture of bone-aſhes contributes to this effect, or cauſes it, upon other principles which are leſs exceptionable. There is no reaſon to doubt that bone-aſhes, whoſe grains are tubulated in all poſſible directions, which greedily imbibe water and emit air, and which render the mortar in which they are mixed manifeſtly bibulous, facilitate the entry of acidulous gas into the cement; and that this matter entering as faſt as the water exhales, occupies the place of the water in the cement, and by preventing the contraction of it, prevents fissures. The speedy induration of the cement, which implies a quick or copious accession of acidulous gas, according to our experience, is a proof of this agency of bone-ashes, as well as an effect deducible from their texture: and from these premises we may easily conceive how they accelerate the setting of calcareous incrustations, and tend to secure them from the injuries of variable weather.

These properties of bone-ashes render them peculiarly useful in incrustations made within doors on principal walls; and the admixture of them in half the quantity of the lime, or in a greater quantity, is the improvement which I pointed out in the twentieth section, whereby the damp, which disfigures the common incrustations made in the circumstances there described, may be obviated, without our incurring the expence of lath-work.

Those who know that one-sixth part of charred bones, or about one-tenth part of

well

well burned bones, is phofphoric acid, may have fome doubt concerning the duration of a cement in which they are mixed in large quantity, unlefs they confider that the ftrength of the cement does not depend on them, and that it is impoffible for the phofphoric acid to quit the lime of bone-afhes, in order to diffolve the faturated lime of the cement. Tho' the bone-afhes fhould perifh in a century, which is not probable, the cement is not likely to fail on this account, provided the quantity of them is not exceffive.

Thus I furmounted the difficulties mentioned in the fourteenth fection, and made my beft calcareous cement applicable in all cementitious and cruftaceous works external or internal, without inducing in it any difagreeable colour or other imperfection.

SECTION XXII.

The Specification made in Consequence of Letters Patent, illustrated with Notes.

IN order to guard against abuses, and to make some compensation for the expences and risques of the artists who publicly and boldly executed, on the great scale, what I had designed; I secured an exclusive right in my cement, by virtue of his majesty's letters patent, on the eighth of January 1779: I authorized Mr. James Wyatt the architect of Queen-Ann-street Cavendish-Square, to use it in the fullest extent, knowing that he, by his knowledge of this subject and his distinguished taste in architecture, will unite in it all the advantages of duration and elegance: I likewise extended this right to Samuel Wyatt the builder in Berwick-street Soho, who is well instructed, and provided with the means of executing any work with this cement, in the highest perfection: And I intend to reserve this priviledge to them, until

til the public convenience requires that it should be extended to others, who are capable of making the same dispositions for the benefit of their employers, and for preserving the reputation of my invention free from the usual exactions of monopolists and the abuses of under-jobbers.

As the specification of these letters patent comprehends the most useful practical instructions deduced from the foregoing experiments and observations, and may serve as a concise recapitulation, I subjoin a transcript of it.

SPECIFICATION.

To all to whom these presents shall come &c.

" Now know ye that in compliance with
" the said provisoe, I the said B. H. do here-
" by declare that my invention of a water
" cement or stucco, for building repairing and
" plastering walls, and for other purposes, is
" described in the manner following (that is
" to say) drift sand, or quarry[1] sand, which

[1] This is commonly called pit-sand.

" consists

" confifts chiefly of hard quartofe flat faced
" grains with fharp angles; which is the
" freeft, or may be moft eafily freed by
" wafhing, from clay, falts, and calcareous
" gypfeous or other grains lefs hard and
" durable than quartz; which contains the
" fmalleft quantity of pyrites or heavy me-
" tallic matter jnfeparable by wafhing; and
" which fuffers the fmalleft diminution of
" its bulk in wafhing in the following
" manner, is to be preferred before any
" other [3]. And where a coarfe and a fine
" fand of this kind, and correfponding in
" the fize of their grains with the coarfe
" and fine fands hereafter defcribed, cannot
" be eafily procured, let fuch fand of the
" foregoing quality be chofen, as may be
" forted and cleanfed in the following
" manner.

" LET the fand be fifted in ftreaming
" clear water, through a fieve which fhall
" give paffage to all fuch grains as do not

---

[2] THE twelfth fection treats of this.

[3] THE reafons of this preference are given in the fifteenth, fixteenth, nineteenth, and twentieth fections.

" exceed

"exceed one sixteenth of an inch in dia-
"meter; and let the stream of water and
"the sifting be regulated so that all the sand
"which is much finer than the Lynn-sand
"commonly used in the London glass-houses,
"together with clay and every other matter
"specifically lighter than sand, may be washed[4]
"away with the stream, whilst the purer and
"coarser sand, which passes thro' the sieve,
"subsides in a convenient receptacle, and
"whilst the coarse rubbish and shingle[5] re-
"main on the sieve, to be rejected.

"Let the sand which thus subsides in
"the receptacle, be washed in clean stream-
"ing water, through a finer sieve, so as to
"be further cleansed and sorted into two
"parcels; a coarser, which will remain in
"the sieve which is to give passage to such
"grains of sand only as are less than one
"thirtieth of an inch in diameter, and
"which is to be saved apart under the name

---

[4] The grounds of this treatment appear in the twelfth and thirteenth section.

[5] I find that I have used this word improperly, on bad authothority. The reader is requested to read rubble instead of shingle throughout this specification.

"of

" of coarse sand; and a finer, which will
" pass through the sieve and subside in the
" water, and which is to be saved apart
" under the name of fine sand.—Let the
" coarse and the fine sand be dried sepa-
" rately, either in the sun, or on a clean iron
" plate set on a convenient furnace, in the
" manner of a sand heat [6].

" LET lime be chosen [7] which is stone lime,
" which heats the most in flaking, and flakes
" the

[6] THE sand ought to be stirred up continually until it is dried, and is then to be taken off; for otherwise the evaporation will be very slow, and the sand which lies next the iron plate, by being overheated, will be discoloured.

[7] THE grounds of the instructions comprized in this paragraph, appear in the second, fourth, fifth and eleventh sections. The preference given to stone lime is founded on the present practice in the burning of lime, and on the closer texture of it, which prevents it from being so soon injured by exposure to the air, as the more spongy chalk lime is; not on the popular notion that stone lime has something in it whereby it excels the best chalk in the cementing properties. The real difference between these will be shewen in the next section.

THE gypsum contained in lime stone remains unaltered or very little altered in the lime, after the burning; but it is not to be expected that clay or martial matter should be found in their native state, in well burned lime; for they concrete or vitrifie with a part of the calcareous earth, and constitute the hard

grains

" the quickeft when duly watered; which
" is the frefheft made and clofeft kept;
" which diffolves in diftilled vinegar with
" the leaft effervefcence, and leaves the
" fmalleft refidue infoluble, and in this re-
" fidue the fmalleft quantity of clay gypfum
" or martial matter.

" LET the lime chofen according to thefe
" important rules, be put in a brafs-wired
" fieve to the quantity of fourteen pounds.
" Let the fieve be finer than either of the
" foregoing; the finer, the better it will be:
" Let the lime be flaked[s] by plunging it in
" a butt

grains or lumps, which remain undiffolved in weak acids, or are feperable from the flaked lime by fifting it immediately through a fieve.

[s] THIS method of impregnating the water with lime is not the only one which may be adopted. It is however preferred before others, becaufe the water clears the fooner in confequence of its being warmed by the flaking lime, and the gypfeous part of the lime does not diffufe itfelf in the water fo freely in this way, as it does when the lime is flaked to fine powder in the common method and is then blended with the water; for the gypfeous part of the lime flakes, at firft, into grains, rather than into fine powder, and will remain on the fieve, after the pure lime has paffed through, long enough to admit of the intended feparation; but when the lime is otherwife flaked, the gypfeous grains have time to flake to a finer powder, and

paffing

"a butt filled with soft water and raising it
"out quickly and suffering it to heat and
"fume, and by repeating this plunging and
"raising alternately and agitating the lime,
"until it be made to pass through the sieve
"into the water; and let the part of the
"lime which does not easily pass through the
"sieve be rejected: and let fresh portions of
"the lime be thus used, until as many [9]
"ounces of lime have passed through the
"sieve, as there are quarts of water in the butt.
"Let the water thus impregnated stand in the
"butt closely covered [10] until it becomes clear;
"and through wooden [11] cocks placed at dif-
"ferent heights in the butt, let the clear liquor

passing through the sieve, dissolve in the water along with the lime. I have imagined that other advantages attended this method of preparing the lime water, but I cannot yet speak of them with precision.

[9] If the water contains no more acidulous gas than is usually found in river or rain water, a fourth part of this quantity of lime, or less, will be sufficient.

[10] The calcareous crust which forms on the surface of the water ought not to be broke, for it assists in excluding the air and preventing the absorption of acidulous gas whereby the lime water is spoiled.

[11] Brass cocks are apt to colour a part of the liquor.

"be

" be drawn off as faſt[12] and as low as the
" lime ſubſides, for uſe. This clear liquor I
" call the cementing liquor[13]. The freer the
" water is from ſaline matter, the better will
" be the cementing liquor made with it.

" LET fifty-ſix pounds of the aforeſaid
" choſen lime be flaked, by gradually ſprinkl-
" ing on it, and eſpecially on the unflaked
" pieces, the cementing liquor, in a cloſe[14]
" clean place. Let the flaked part be im-
                                    " mediately

[12] LIME water cannot be kept many days unimpaired, in any veſſels that are not perfectly air-light. If the liquor be drawn off before it clears, it will contain whiting, which is injurious; and if it be not inſtantly uſed, after it is drawn limpid from the butt into open veſſels, it will grow turbid again, and depoſite the lime changed to whiting by the gas abſorbed from the air. The calcareous matter which ſubſides in the butt, reſembles whiting the more nearly, as the lime has been more ſparingly employed; in the contrary circumſtances, it approaches to the nature of lime; and in the intermediate ſtate, it is fit for the common compoſition of the plaiſterers for inſide ſtucco.

[13] AT the time of writing this ſpecification I preferred this term before that of lime-water, on grounds which I had not ſufficiently examined.

[14] THE vapour which ariſes in the flaking of the lime contributes greatly to the flaking of theſe pieces which lie in its way; and an unneceſſary waſte of the liquor is prevented, by applying it to the lime heaped in a pit or in a veſſel which may
                                                    reſtrain

" mediately [15] sifted through the last men-
" tioned fine brass-wired sieve: Let the lime
" which passes be used instantly or kept in
" air-tight vessels, and let the part of the lime
" which does not pass through the sieve, be
" rejected [16].—This finer richer part of the
" lime which passes through the sieve, I call
" purified lime.

" LET bone-ash be prepared [17] in the usual
" manner by grinding the whitest burnt
" bones, but let it be sifted to be much finer

restrain the issue of the vapour, and direct it through the mass. If more of the liquor be used than is necessary to slake the lime, it will create error in weighing the slaked powder, and will prevent a part of it from passing freely through the sieve. The liquid is therefore to be used sparingly, and the lime which has escaped its action is to be sprinkled apart with fresh liquor.

[15] WHEN the aggregation of the lumps of lime is thus broken, it is impared much sooner than it is in the former state, because the air more freely pervades it. This is shewn in the fifth section.

[16] BECAUSE it consists of heterogeneous matter, or of ill burnt lime; which last will flake and pass through the sieve, if the lime be not immediately sifted after the flaking, agreeable to the text. The reason of this may be drawn from the fourth section.

[17] THIS art is taught in the twenty-second section.

" than

" than the bone-ash commonly sold for
" making cupels.

" THE most eligible materials for making
" my cement being thus prepared: Take
" fifty-six pounds of the coarse sand and
" forty-two pounds of the fine sand; mix
" them on a large plank of hard wood
" placed horizontally; then spread the sand
" so that it may stand to the height of six
" inches with a flat surface on the plank;
" wet it with the cementing liquor; and let
" any superfluous[18] quantity of the liquor,
" which the sand in the condition described
" cannot retain, flow away off the plank.
" To the wetted sand add fourteen pounds
" of the purified lime in several successive
" portions, mixing and beating them up toge-
" ther in the mean time with the instruments
" generally used in making fine mortar:
" then add fourteen pounds of the bone-ash
" in successive portions, mixing and beating
" all together. The quicker and the more
" perfectly these materials are mixed and
" beaten together, and the sooner the ce-

[18] THE grounds of this practice are shewn in the twelfth section.

O " ment

"ment thus formed is ufed, the better [19] it
"will be. This I call the water cement
"coarfe grained, which is to be applied in
"building, pointing, plaiftering, ftuccoing,
"or other work, as mortar and ftucco now
"are; with this difference chiefly, that as
"this cement is fhorter than mortar or com-
"mon ftucco and dries fooner, it ought to
"be worked expeditioufly in all cafes, and
"in ftuccoing it ought to be laid on by flid-
"ing the trowel upwards on it; that the
"materials ufed along with this cement
"in building, or the ground on which it is
"to be laid in ftuccoing, ought to be well
"wetted [20] with the cementing liquor, in

[19] These proportions are intended for a cement made with fharp fand, for incruftation in expofed fituations, where it is neceffary to guard againft the effects of hot weather and rain. In general half this quantity of bone-afhes will be found fufficient; and altho the incruftation in this latter cafe will not harden deeply fo foon, it will be ultimately ftronger provided the weather be favourable.

The injuries which lime and mortar fuftain, by expofure to the air, before the cement is finally placed in a quiefcent ftate, appear in many parts of the foregoing pages; and therefore our cement is the worfe for being long beaten, but the better as it is quickly beaten untill the mixture is effected, and no longer.

[20] See fection vii. and page 75.

"the

" the inftant of laying on the cement; and
" that the cementing liquor is to be ufed when
" it is neceffary to moiften the cement, or
" when a liquid is required to facilitate the
" floating of the cement.

" WHEN fuch cement is required to be
" of a finer texture; take ninety-eight pounds
" of the fine fand, wet it with the cementing
" liquor and mix it with the purified lime
" and the bone-afh in the quantities and in
" the manner above defcribed, with this
" difference only, that fifteen pounds of
" lime, or[21] thereabouts, are to be ufed inftead
" of fourteen pounds, if the greater part of
" the fand be as fine as Lynn fand. This I
" call water cement fine grained. It is to
" be ufed in giving the laft coating or the fi-
" nifh to any work intended to imitate the
" finer grained ftones or ftucco. But it may
" be applied to all the ufes of the water ce-
" ment coarfe grained, and in the fame
" manner.

[21] SEE fection xiii. The quantity of bone afhes is not to be increafed with that of the lime, for the reafon given in page 176; but it is to be leffened as the expofure and purpofes of the work will admit. See fection xxii.

"When for any of the foregoing pur-
"poses of pointing, building, &c. such a
"cement is required much cheaper and
"coarser grained, then, much coarser clean
"sand than the foregoing coarse sand, or well
"washed fine [22] shingle is to be provided.
"Of this coarsest sand or shingle [22] take fif-
"ty-six pounds, of the foregoing coarse sand
"twenty-eight pounds and of the fine sand
"fourteen pounds; and after mixing these
"and wetting them with the cementing li-
"quor in the foregoing manner, add 14
"pounds, or somewhat less, of the [23] puri-
"fied lime, and then fourteen pounds or
"somewhat less of the bone-ash, mixing
"them together in the manner already de-
"scribed. When my cement is required
"to be white, white sand, white lime, and
"the whitest bone-ash are to be chosen.
"Grey sand and grey bone-ash formed of
"half burnt bones, are to be chosen to make
"the cement grey; and any other [24] colour

---

[22] Rubble.

[23] BECAUSE less lime is necessary as the sand is coarser. Section xii. and xiii.

[24] THE outlines of these arts are given in section xx.

"of

" of the cement is obtained, either by chuf-
" ing coloured fand, or by the admixture of
" the neceſſary quantity of coloured talc in
" powder, or of coloured vitreous or metal-
" lic powders, or other durable colouring
" ingredients commonly uſed in paint.

" To the end that ſuch a water cement
" as I have deſcribed may be made as uſe-
" ful as is poſſible in all circumſtances; and
" that no perſon may imagine that my
" claim and right under theſe Letters Pa-
" tent may be eluded by divers variations
" which may be made in the foregoing pro-
" ceſs without producing any notable de-
" fect in the cement; and to the end that
" the principles of this art as well as the art
" itſelf of making my cement, may be ga-
" thered from this ſpecification and perpe-
" tuated to the public, I ſhall add the fol-
" lowing obſervations.

[21] THE known chemical properties of the ſeveral finer ingredients uſed in paint or water colouring, and the experienced
effects of the materials mentioned in the fifteenth, ſixteenth,
ſeventeenth, eighteenth, nineteenth and twentieth ſections,
are ſufficient to direct the artiſt in the choice of thoſe things
which will induce colour, with the ſmalleſt injury to the
incruſtation.

"This my water cement, whether the coarse or fine grained, is applicable in forming artificial stone, by making alternate layers of the cement and of flint, hard stone, or brick, in moulds of the figure of the intended stone, and by exposing the masses so formed, to the open [26] air to harden.

"When such cement is required for water [27] fences, two thirds of the prescribed

---

[26] But they must not be exposed to the rain, until they are almost as strong as fresh Portland stone; and even then they ought to be sheltered from it, as much as the circumstances will admit. See pages 68, 69, 114. These stones may be made very hard and beautiful, with a small expence of bone-ash, by soaking them, after they have dried thoroughly and hardened, in the lime-liquor, and repeating this process twice or thrice, at distant intervals of time. The like effect was experienced in incrustations, and is mentioned in page 114.

[27] To what I have said on this subject in page 124, I must add that, in my experiments, mortar made with terras powder, in the usual method, does not appear to form so strong a cement for water fences, as that made according to the specification, with coarse sand; and I see no more reason for avoiding the use of sand in terras mortar, than there would be for rejecting stone from the embankment. The bone-ashes meant in this place are the dark grey or black sort. I am not yet fully satisfied about the operation of them in this instance.

"quantity

" quantity of bone afhes are to be omitted;
" and in the place thereof an equal meafure
" of powdered terras is to be ufed; and if
" the fand employed be not of the coarfeft
" fort, more terras muft be added, fo that
" the terras fhall be by weight one fixth part
" of the weight of the fand.

" WHEN fuch a cement is required of the
" fineft grain[23] or in a fluid form, fo that
" it may be applied with a brufh, flint pow-
" der, or the powder of any quartofe or hard
" earthy fubftance may be ufed in the place
" of fand, but in a quantity fmaller as the
" flint or other powder is finer; fo that the

---

[23] THE qualities and ufes of fuch fine calcareous cement are fet forth in the thirteenth and twentieth fections. They are recommended chiefly for the purpofe of fmoothing and finifhing the ftronger cruftaceous works, or for wafhing walls to a lively and uniform colour. For this laft intention, the mixture muft be as thin as new cream, and laid on brifkly with a brufh, in dry weather; and a thick and durable coat is to be made by repeated wafhing, but is not to be attempted by ufing a thicker liquor; for the coat made with this laft is apt to fcale, whilft the former endures the weather much longer than any other thin calcareous covering that has been applied in this way. Fine yellow ochre is the cheapeft colouring ingredient for fuch a wafh, when it is required to imitate Bath ftone, or the warm-white ftones.

" flint powder or other such powder shall not
" be more than six times the weight of the
" lime, nor less than four times its weight.
" The greater the quantity of lime within
" these limits, the more will the cement be
" liable to crack by quick drying, and vice
" versa.

" WHERE such sand as I prefer cannot be
" conveniently procured, or where sand can-
" not be conveniently washed and sorted,
" that sand which most resembles the mix-
" ture of coarse and fine sand above pre-
" scribed, may be used as I have directed,
" provided due attention is paid to the quan-
" tity of the lime, which is to be the
" greater ⁹ as the sand is the finer and vice
" versa.

" WHERE sand cannot be easily pro-
" cured, any durable stoney body, or baked
" earth grosly powdered¹⁰ and sorted nearly
" to

⁹ FURTHER instructions may be gathered from the thirteenth section. If sea sand be well washed in fresh water, it is as good as any other *round* sand.

¹⁰ THE cement made with these and the proper quantitie of purified lime and lime-water, are inferior to the best, as

the

" to the fizes above prefcribed for fand, may
" be ufed in the place of fand, meafure for
" meafure, but not weight for weight, un-
" lefs fuch grofs powder be as heavy fpe-
" cifically as fand.

" SAND may be cleanfed from every fofter
" lighter and lefs durable matter and from
" that part of the fand which is too fine, by
" various methods preferable [1] in certain
" circumftances, to that which I have de-
" fcribed.

" WATER may be found naturally free from
" fixable gas felenite or clay: fuch water may,
" without any notable inconvenience, be ufed
" in the place of the cementing liquor; and
" water approaching this ftate will not re-
" quire fo much lime as I have ordered, to

the grains of thefe powders are more perifhable and brittle than thofe of fand. They will not therefore be employed, unlefs for the fake of evafion, or for want of fand: in this latter cafe the finer powder ought to be wafhed away.

[2] THIS and the next paragraph is inferted with a view to evafions, as well as to fuggeft the eafier and cheaper methods which may be adopted in certain circumftances, by artifts who underftand the principles which I have endeavoured to teach.

" make

"make the cementing liquor; and a cement-
"ing liquor sufficiently useful may be made
"by various methods of mixing lime and
"water in the described proportions, or
"nearly so.

"When stone lime cannot be procured,
"chalk lime or shell lime which best re-
"sembles stone lime, in the characters above
"written of lime, may be used in the manner
"described, except that [32] fourteen pounds
"and a half of chalk lime will be required in
"the place of fourteen pounds of stone lime.
"The proportion of lime which I have pre-
"scribed above may be encreased without
"inconvenience, when the cement or stucco
"is to be applied where it is not liable to
"dry quickly; and in the contrary circum-
"stance this proportion may be diminished;
"and the defect of lime in quantity or
"quality may be very advantageously sup-

---

[32] This relates to chalk lime burned with a sufficient quantity of fuel in kilns of the common construction. Chalk lime prepared as I shall shew in the next section, will go as far as stone lime, if not farther.

"plied

" plied ", by caufing a confiderable quantity
" of the cementing liquor to foak into the
" work, in fucceffive portions and at diftant
" intervals of time, fo that the calcareous
" matter of the cementing liquor, and the
" matter attracted from the open air, may
" fill and ftrengthen the work.

" The powder of almoft every well
" dried or burnt animal fubftance may be
" ufed inftead of bone-afh; and feveral
" earthy powders, efpecially the micaceous
" and the metallic; and the elixated afhes
" of divers vegetables whofe earth will not
" burn to lime; and the afhes of mineral
" fuel, which are of the calcareous kind, but
" will not burn to lime; will anfwer the ends
" of bone-afh in fome degree [34].

---

[33] This practice is noticed, as the remedy which may be ufed for the defects arifing from evafive meafures, and as the method of giving fpongy incruftations containing bone-afhes, the greateft degree of hardnefs.

[34] The ufeful fubftitutes for bone-afhes, have been treated of in the foregoing fections: the metallic micaceous and earthy powders are not recommended in the text, but only enumerated for reafons which influenced the ftyle of this fpecification, and which lawyers will perceive.

" The

" The quantity of bone-ash described may be lessened without injuring the cement, in those circumstances especially which admit the quantity of lime to be lessened, and in those wherein the cement is not liable to dry quickly. And the art of remedying the defects of lime may be advantageously practised to supply the deficiency of bone-ash, especially in building and in making artificial stone with this cement.

" N. B. For inside work, the admixture of hair with this cement is useful.
" In witness whereof I the said B. H. &c."

The excellence of my cement depends first, on the figure size and purity of the sand; secondly on the purity of the lime, obtained in the choice of lime-stone, and in the perfect burning, and secured in the preservation of it from air, in my method of slaking, and in the separation of heterogeneous parts; thirdly on the use of strong and pure lime water in the place of common water; fourthly on the proportion of sands lime water and lime; fifthly, on the manner of mixing them; sixthly, on the knowledge of ingredients

dients and circumftances which are injurious or ufeful; feventhly, on the ufe of bone-afhes of determinate fize; eighthly, on the art of fuiting fome of thefe to the feveral purpofes; and finally on fo many other particulars, as render it very difficult to give a more candid fpecification, in the ufual compafs, than this which I have enrolled, or to guard otherwife againft evafions, than by anticipating them.

I do not think it neceffary to infift more minutely on the mechanical arts of applying the coarfer or finer calcareous cement, to produce the moft agreeable effects, becaufe they are known to fo many workmen employed under Meff. Wyatt, and are fo nearly related to thofe already known to the plaifterers, that they are not likely to be miffed or loft.

SECTION

SECTION XXIV.

Experimental Comparisons of Chalk-Lime with Stone-Lime. Advices to the Manufacturers of Chalk-Lime, concerning the Art of rendering it equal, if not superior, to Stone-Lime, for the Purposes of Builders Soap-Boilers and Sugar-Bakers.

ALL the authors whom I have consulted, who have treated of cementitious buildings and of lime, from the time of Vitruvius, who wrote on these subjects in the reign of the Roman Emperor Titus or before it, down to the present hour; and all the artists with whom I have conversed, agree in the opinion that lime prepared from the closest lime-stone makes a stronger cement than that which is made of spongy lime-stone, and that the lime of chalk particularly, is incapable of acting as effectually as the best stone lime, in cementitious works or incrustations which are exposed to the weather.

THIS universal and unquestioned notion had great influence with me in the course of my experiments, until I had discovered not only the fallacy of it, but the grounds which gave rise to it: both which I shall now expose, in the pleasing hope of rendering great services to many of my friends, and all who are proprietors of chalk-pits, or are obliged to use chalk-lime in their buildings.

THE experiments already mentioned afforded me a great many opportunities of comparing cements made with lime and sand, or with these and other ingredients in various proportions, and differing only in the kind of lime. In these comparisons I could not perceive that chalk lime, judiciously prepared and used, was in any respect inferior to the best stone lime: but I did not content myself with these. I made a great number of cements, with the sole view of collating the respective merits of these kinds of lime, in small and great incrustations, in masses made to resemble cut stone, in all exposures and seasons of the year; and after the strictest comparisons of those which contained lime in equal quantities and were treated

alike

alike in all refpects, I was thoroughly convinced that my chalk lime was as good for any purpofe of this kind, as the beft ftone lime in this kingdom; for I ufed the well-burned lime of Plymouth ftone, which I reckon among the moft excellent of our lime ftones.

PLYMOUTH lime-ftone lofes feven fixteenths of its weight, in the converfion to lime, and becomes as white as chalk. Chalk loofes a little more in the perfect burning. Plymouth lime leaves a fmall gypfeous refidue in the folution prefcribed in the tenth page, which is preferable to that directed in the fpecification: chalk-lime leaves none. Therefore the chalk-lime chemically or technically tried, appears to be equal, if not fuperior to ftone lime, in its cementing powers, when it is properly ufed.

THE prejudices entertained againft chalk-lime may be traced to three fources. The firft is that which is mentioned in the fourth fection. The vulgar criterion of the due preparation of lime confifts in the flaking: and as chalk, which has undergone a flight calcination

nation and thereby loſt only a part of its acidulous gas, is capable of flaking, by reaſon of its ſponginefs; the manufacturers of chalk lime content themſelves with the degree of calcination which renders it tractable or vendible, and thus bring it into diſrepute.

THE ſecond ſource is mentioned in the fifth ſection. Chalk lime imbibes acidulous gas, during its expoſure to the air, much faſter than ſtone lime, and is conſequently more impaired or worſe, at the time of uſing it in mortar, than ſtone lime kept in the ſame circumſtances. As the lime may be greatly injured in this way, without flaking ſenſibly; and as there was no ſuſpicion or meaſure of ſuch injury, beyond what the flaking afforded; the acquired imperfection of chalk lime was conſidered as the very nature of it. In the thirtieth page it appears that a pound of chalk lime, placed in the quieſcent air of a chamber, imbibes two ounces and a half of acidulous gas in two days, which is the ſhorteſt time in which lime is uſually expoſed, if we count from the moment of its being red hot to that of its being mixed in mortar, during which interval it is in the ſtate of abſorption.

P

The third source I have discovered in the structure of lime kilns. The cavity of a lime-kiln has the figure of a truncated cone inverted. When the charge, consisting of lime stone and fuel alternately stratified, has burned for some time, the fuel is exhausted at the lower narrow extremity of the cavity, the lime in this part cools, and serves as a grate to the fuel and limestone above it, which continue to burn briskly, for eighteen hours or longer, after the lime beneath begins to cool. During this time the last-mentioned part of the lime is exposed to a strong current of air; and the whole charge of lime stands in the like current of air until the kiln is cooled, or the lime is withdrawn; which in common practice is seldom or never done before the sixtieth or seventieth hour after the combustion of the fuel commenced.

The injury which lime stone sustains, in these circumstances, which I have often imitated in my elaboratory, is not great; because this lime is much more compact than the chalk lime. But when we observe that the best pieces of chalk lime of the common kilns,

and

and thofe, which, after heating them fufficiently, I had left in the fire-place, expofed as they are in the ufual procefs, are always effervefcent; that good chalk lime, in a weaker current of air, imbibes more than three ounces of acidulous gas into each pound of it, in two days, according to the experiment of the thirtieth fection; and that my chalk lime, which I remove from the fire-place as foon as it is fufficiently burned, is perfectly non effervefcent; we find that the long experienced imperfections of frefh chalk lime are owing more, to the faulty conftruction of the kilns, and the ignorance of the manufacturer, than to any incapacity of chalk to yield excellent lime.

The means of preparing chalk lime to equal or exceed ftone lime, and of making the beft ftone lime, may be gathered from what I have faid, and the following intimations. The kilns are to be made broader and fhallower in the cavity which receives the charge: the circular wall inclofing this cavity, is to be continued tapering upwards, until it terminates in a lofty flue, in order to accelerate the combuftion and increafe the heat by a quick current of air to be regulated by opening or closing the door-way, which is to be left in the circular wall at a convenient height for the introduction of the charge. The massive walls of the lower cone are to be lined with fire brick or apyrous stone set in the best fire loam; and are to be girded with iron. The fuel is to be so stratified with the lime stone or chalk, and the combustion is to be so conducted, that every part of the charge shall be sufficiently ventilated and heated, and that the lowest shall remain red hot until the whole is well burned. Then the current of air through the kiln is to be stopped, by closing the apertures at the bottom; or the red hot lime is to be removed out of it, to cool in quiescent air, until it is fit to be inclosed in air-tight vessels. A cask of chalk lime is not to be opened until the moment when the workman is ready to slake the lime; and the greatest expedition is to be used in the slaking, in making the mortar, and in applying it to use. By this treatment the chalk lime will answer every end of the best stone lime; and stone lime may be prepared and preserved in the highest perfection which the nature of the lime stone can admit.

THE

The manufacturer of chalk lime who first adopts these measures near London will profit by them: for good lime is not only better, but goes farther in building, than bad lime; good chalk lime will answer the purposes of the soap-boiler, in half the quantity which they use of common chalk lime, the greater part of which serves only to waste their lees and clog their vatts; and our sugar-bakers will not hesitate about the price of good chalk lime, when they find that it is totally soluble in pure water, and introduces no selenitic matter into the sugar. The exportation of lime to the West India Islands will be a further incitement towards the improvements which I have suggested, when the planters receive the information which I intend soon to give them, concerning the principles by which *lime, duly administered facilitates, but injudiciously used impedes the granulation* of the saccharine part of their cane juice.

SECTION XXV.

Directions to the Houses already stuccoed with the new Cement. Observations on the Objections of certain Artists; on the cementitious Works of the Romans; on the experienced and unequalled Duration of such Cements; on the Cements of Loriot and others; and on certain Uses of the Author's Cement.

THE inexperience of the workmen, their obstinate adherence to their own notions, and the opinion which they entertained that some of the rules prescribed to them were insisted on rather through an affectation of mystery than for any useful purpose, operated strongly against the best endeavours of Messieurs Wyatt, in the incrustations first made on the great scale for use or ornament. In consequence of these disadvantages, which will be obviated in future, their stucco, although it excels others beyond comparison and is far from being perishable, is not quite so hard as it might have been made. This I mention, lest these incrustations should

be

be mistaken for the best, which I have represented as exceeding Portland stone in hardness. These last demand a strict observance of the foregoing precepts respecting the season and the exposure as well as the materials and mechanical application of them.

The houses which have been stuccoed with this cement are the following:

The north front of Mr. Delme's house on the south side of Grosvenor-Square, stuccoed in November and December 1778.

The north and south fronts of Mr. Viner's house in Conduit-Street, Hanover-Square, and the moulded walls of the area behind it, stuccoed in the summer and autumn of 1779: the fore-front representing Bath stone, the other front and the walls and mouldings of the area, closely imitating Portland stone.

Mr. Bond Hopkins's house at Wimbledon in Surry, stuccoed in the summer of 1779, in every aspect.

Mr. Birch's houfe at Hamftead near Birmingham, ftuccoed in autumn 1779, in every afpect.

These were done under the direction of Mr. James Wyatt.

I must obferve that the difcoloured fummit of the front of Mr. Viner's houfe, is natural ftone; and that our cement never changes its colour, like thofe which contain white lead or oil. This houfe feems to be well finifhed. But when an incruftation is made in an improper feafon; when the parapet or gutters are defective, and the rain is fuffered to penetrate through the fpongy bricks and recent ftucco; it is the fault of the workman and not of the cement, if the damp appears in the incruftation of the attic ftory, or if this part of it fhould never harden compleatly.

I have not yet been nformed of the work done by Mr. Samuel Wyatt, except the ftuccoing of the Honourable Juftice Willes's houfe at Little Grove Eaft Barnet, on the northern fouthern and eaftern fides, in September and October 1779; and a piece of incruftation reprefenting a very coarfe ftone, made in November 1778, from the foundation to the height of the cellar ftory, on the eaftern wall of Mr. Curzon's houfe in Davies-Street Grofvenor-Square. The earth had lain againſt this wall for many years: it was ftuccoed immediately after the area had been opened to it, and whilft it was damp; and by the miſtake of the painter, the ftucco was painted whilſt it was freſh: I have not feen it fince: but I am told that under all thefe difadvantages it is incomparably better than the piece of common ftucco which meets it from above.

WHAT has been fairly ſhewn of the cement, in this public manner, has given greater fatisfaction; and Meffrs. Wyatt are engaged to ftucco a great number of capital houfes with it next fummer. Thefe will be done in the higheft perfection, becaufe the workmen are now compliant and experienced.

AN impediment however ſtill fubfifts to obftruct the progrefs of this art in the public eftimation. Some interefted perfons diligently infinuate that this cement has not the fanction of long experience; and that however promifing it now appears, it may moulder like others in a few years hence: they likewife obferve that only a fuperficial cruſt of
the

the above-mentioned ſtuccoes is ſufficiently hard, but that the internal parts may eaſily be cut or broke: and they repreſent the cement as an expenſive compoſition.

This publication, which ſhews it to be made of clean ſorted ſands, of bone-aſhes, and of good lime in about half the quantity uſed in the common method of making mortar, renders an anſwer to the laſt-mentioned objection quite unneceſſary.

The ſecond objection may be thus conſidered. The ſame attractive power, which draws acidulous gas from the air into lime, muſt neceſſarily prevent any conſiderable quantity of the gas from entering deeply in a recent incruſtation, until the lime at the ſurface is ſaturated with it, and conſequently until a ſuperficial layer of the cement is highly indurated. The ſame matter, which during its acceſſion hardens this part, muſt likewiſe tend to render it cloſer in the texture and leſs freely pervious to that which is ſtill neceſſary for the induration of the internal portions: and thus it happens that an incruſtation grows harder at the ſurface, in one

week,

week, than it does deeply in the substance, in a year, although bone-ashes be used to lessen the obstruction of the surface. Since therefore we know the reason why the interior parts of an incrustation cannot harden in a much longer time than is necessary for the hardening of the exterior; since the materials of the cement are the same in the central parts as well as at the superfices, and must be equally affected by acidulous gas when it can reach to them; since by our experience of old cements composed of lime and sand, we know that the induration extends equally through the mass of them, in the course of years; it is manifest that our stucco will harden in due time through the whole substance of it, as much as it does in a shorter time at the surface, and that the objection, founded on the internal weakness noticed in the first year, is futile.

To prove this by experiment, scrape away the hardened superficial stratum, as I have often done; and taking care to brush off all that you have loosened beneath, leave the new surface of the friable part of the cement exposed to the air for a few weeks. You will find

find it to harden like the firſt ſurface, whilſt the parts beneath it ſtill continue brittle. You will perceive after a repetition of the like experiment in the ſame place, that every part of the ſtucco is capable of acquiring the hardneſs of the firſt ſurface, in a few days, and conſequently that the whole will accquire it in the longer time neceſſary for the entrance of acidulous gas through the compact exterior cruſt.

WITH regard to the objections grounded on our ſhort experience of this cement, I think they can have very little influence amongſt informed men who know, from the writings of the antients, by the inſpection of old cements, and by the analyſis of them, that mortar made of lime and ſand can endure every trial of the weather in the moſt expoſed ſituations for a thouſand years or more. Such objections deſerve no better anſwer than ought to be given to an illiterate London bricklayer, who ſhould object to the uſe of porphyry in building, becauſe he has no certainty of its being ſo durable as the

bricks

bricks which he had for many years experienced.

I AM aware of the opinion, which is prevalent at this time, that the antients used something which is unknown to us in their mortar, and that this long lost ingredient is the cause of the duration and hardness of those cements which we so much admire in some of their structures. A notion founded on conjecture does not demand a serious discussion. I will therefore treat it as a subject of conversation rather than of argument.

VITRUVIUS in the fifth chapter of the second book of his architecture speaks thus of lime.

Quare autem cum recipit aquam, & arenam calx, tunc confirmat structuram, hæc esse causa videtur, quod e principiis uti cætera corpora, ita & saxa sunt temperata: & quæ plus habent aeris, sunt tenera: quæ aquæ, lenta sunt ab humore: quæ terræ, dura: quæ ignis, fragiliora. Itaque ex his saxa, si antequam coquantur, contusa minute, mixtaque arenæ conjiciantur in structuram, nec solidescunt, nec eam poterunt continere: cum vero

conjecta

conjecta in fornacem, ignis vehementi fervore correpta, amiserint pristinæ soliditatis virtutem, tunc exustis, atque exhaustis eorum viribus, relinquuntur patentibus foraminibus, & inanibus: ergo liquor, qui est in ejus lapidis corpore, & aer cum exhaustus, & ereptus fuerit, habueritque in se residuum calorem latentem, intinctus in aqua prius, quam exeat ignis, vim recipit, & humore penetrante in foraminum raritates confervescit, & ita refrigeratus rejicit ex calcis corpore fervorem. Ideo autem quo pondere saxa conjiciuntur in fornacem, cum eximuntur, non possunt ad id respondere, sed cum expenduntur, eadem magnitudine permanente, excocto liquore circiter tertia parte ponderis imminuta esse inveniuntur. Igitur cum patent foramina eorum, & raritates, arenæ mixtionem in se corripiunt, & ita cohærescunt, siccescendoque cum cæmenis coeunt, & efficiunt structurarum soliditatem.

The same ignorance of the nature of lime is betrayed by Alberti and later writers. And since we do not find any scientific rules prescribed by literary artists, for the composition of calcareous cements with such chosen and sorted materials as I have described, or in such proportions of them; and since it is highly improbable that the remembrance

membrance of an ufeful ingredient, or any knowledge once accquired in an art practifed in fo many countries nd by fo many different perfons in every age, fhould have been loft; we have the moft fatifactory reafons for concluding that the antients had no fkill beyond that of our modern builders, in the preparation of lime or mortar.

THE ruins of Herculaneum, and other reliques of their works, furnifh us with abundance of bad mortar and defective incruftations, which are inftances of their ignorance of thofe principles by which the beft cement might be made equally cheap. The total ruin and obliteration of many of their buildings, argue to the fame end; for well cemented works fuffer very little by dilapidation, by reafon of the difficulty and expence of pulling them to pieces and applying the materials to other ftructures. If to thefe confiderations I can add an expofition of the fortuitous circumftances which rendered fome of their cements uncommonly hard and durable, I hope I fhall not be fufpected of ungenerous invidious motives, in faying that the aqueducts and other ftructures,

which

which have been preserved to us through so many ages, by the strength of their cement, are monuments rather of the good luck, than of any extraordinary skill, of those who built them.

When the neighbouring quarries afforded good lime stone, free from gypsum, and such as required to be well burned before it could slake freely; when the preparation of the lime, at the public expence, afforded no temptation for parsimony in fuel; and when the vicinity of the lime stone, and the quick consumption of the lime in great massive works, prevented those injuries which it sustains in long transportation and exposure, in the slaking of great quantities of it at once, or in the keeping of mortar made with it; the ignorance of the artists could not produce any defects dependent on bad lime, because necessity or chance inforced all that could have been sought by choice, in this instance.

When the vicinity afforded sand, clean quartose sharp well sized and resembling our mixture of the coarse and fine; chance fur-
nished

nifhed all that fkill could aim at, in the choice and preparation of this article.

When walls of immenfe thicknefs were conftructed chiefly with fmall ftones, in the way of boulder-work, the great confumption of mortar made every practicable faving of lime an object of great importance; and as the mortar muft be made ftiff for fuch work, it was neither convenient nor neceffary to mix much lime in it, or to ufe fine fand in it, or to exclude the rubble from it: and thus, by motives of œconomy and convenience, rather than by any others, they were led to the meafures which infured to the cement of fuch ftructures every perfection dependent on the goodnefs of lime and fand and on good, if not the beft, proportions of them.

When the ftones ufed in building were recently dug, or collected from the beds of rivers, the artifts needed no precautions againft the bad effects of dry bibulous and dufty ftones or bricks; and their works had, of neceffity, every good quality attainable by the practice, which I commend, of foaking thefe materials. When their water was good, the cement abounding in lime was not much the worse for their ignorance of the use of lime water.

When the structure was intended to stand by its own strength, rather than to depend on timbers; and was by the solidity of its bearings and the diameter of its stoney substance, secured from agitation; when the thickness of the walls, prevented the cement from being hastily dried, and afterwards secured it from being thoroughly wetted; and when the enormous weight contributed to the approximation and cohesion of the parts of the cement to each other and to the stones; every defect of cementitious buildings, of a contrary description, was obviated by the nature of the structure, which rendered it as perfect, in the hands of any artists, as the most consummate skill could make our modern slender tremulous bibulous walls.

In the concurrence of these circumstances, we find excellent cements of great antiquity, which I need not point out to literary men: but since they are found no where else, that I have discovered; and since it is not probable that the antients had any art of this kind un-

known

known to the moderns, I think I am authorized to conclude that their beft cementitious works, inftead of being held forth as inftances of their unequalled fkill, ought rather to be confidered as fubftantial proofs of the duration of mortar or ftucco duly compofed of fand and lime, beyond all others, and of the utility of thefe endeavours which I have made for preparing calcareous cements according to fcientific principles, which enable us to make them in the higheft perfection in all places, and to accommodate them to every purpofe of ufe or ornament.

I have ftudioufly avoided ftrict comparifons of my cement or of the beft Roman cements, with the oil cements; becaufe the beft of thefe is private property; not doubting that my liberal readers will give this filence a conftruction equally favourable to the proprietors and to me. But it is not neceffary to lay myfelf under the fame reftraint refpecting the reputed improvement of Monfieur Loriot, publifhed in 1774 at Paris, in a pamphlet entitled, " A treatife on a new difcovery in the " art of building, made by Monfieur Loriot, " mechanic and penfionary to the king; in " which

"which is announced, by order of his ma-
"jefty, the method of compofing a cement
"or mortar fit for an infinity of works as
"well in building as in decoration."

The firft half of this effay ferves only to difplay the fanguine hopes and lively imagination of the author, which tranfported him beyond the bounds of his knowledge in this fubject, and all the rules of phyfical induction. In the thirty-firft page he fays that "the admixture of powdered quicklime, in "any mortar made whith flaked lime, is "the moft effectual method of giving it "every defirable perfection; and that this "is the chief difcovery which he announ- "ces." In the next page he gives the following prefcription.

"Take one part of brick-duft finely fift-
"ed, two parts of fine river fand fkreened,
"and as much old flaked lime as may be
"fufficient to form mortar with water, in
"the ufual method, but fo wet withal as to
"ferve for the flaking of as much powdered
"quick-lime as amounts to one fourth of the
"whole quantity of brick-duft and fand.
 "When

" When the materials are well mixed, em-
" ploy the compofition quickly, as the
" fmalleft delay may render the application
" of it imperfect or impoffible."

In the 40th page he fays, " Another me-
" thod of making the compofition is, to
" make a mixture of the dry materials;
" that is to fay, of the fand brick-duft and
" powdered quick-lime, in the prefcribed
" proportion; which mixture may be put
" in facks, each containing a quantity fuf-
" ficient for one or two troughs of mortar.
" The abovementioned old flaked lime and
" water being prepared apart, the mixture
" is to be made, in the manner of plaifter,
" in the inftant when it is wanted, and even
" on the fcaffold, and is to be well chafed
" with the trowel."

To exprefs Mr. Loriot's difcovery briefly and difpaffionately, I would fay, when an ignorant artift makes mortar with whiting inftead of lime, he can mend it confiderably by adding lime to it: but his mortar will ftill be defective, in comparifon with the beft that may be made, by reafon of the old flaked lime or whiting. For on repeated trials I found this to be the true state of the case.

Lime sustains less injury in powdering small quantities of it in a covered vessel, than by flaking, in the usual method, with common water: and powdered lime, in the mixing of it with sand and water, excites a warmth in the mass, which greatly contributes to its drying or setting quickly. Mr. Loriot not knowing how to flake lime without impairing its cementing virtue, and taking the speedy exsiccation as an omen of perfect induration, imagined the prescription of powdered lime to be a great improvement.

If the powdering of lime, without exposing it much to the air, were not an expensive operation, I should have directed this powder to be used instead of water-flaked lime, for those parts of an *incrustation*, which are prevented from drying in due time, by their vicinity to the damp earth or to projections on which the rain lodges. The cement to be applied in such circumstances ought,

ought, as I said in the specification, page 202, to contain more lime than I have prescribed for other situations; and it will be found the better for being made with powdered quick-lime, because it will dry the sooner, and becoming pervious to air in consequence of the exhalation of its water, it will more speedily acquire that hardness which secures it from being exhausted of the lime by the constant moisture or the trickling rain.

The public are indebted to Mr. Hartley for the experimental proofs he has given of the efficacy of his method of securing houses from fire; and to lord Mahon for those judicious and expensive experiments by which he has shewn that a calcareous incrustation answers the purposes of Mr. Hartley's art. I am afraid that their good intentions will be frustrated by the indifference of men to distant or improbable evils, and their dislike to any immediate expence which affords no extemporary convenience or ornament. But altho' such motives of œconomy should dissuade us from adopting their measures in the fullest extent, we ought certainly to avail
ourselves

ourselves of the useful knowledge which they have imparted, so far as to prefer a safe and durable stucco, wherever it is applicable by the assistance of hair, before wainscot or wooden ornaments. For although no metallic or calcareous covering can secure the wood of a house from being charred by a great fire, the danger of others is lessened as the combustible materials are secured from the action of the air, and consequently from contributing to the deflagration.

I have thought that the small stones, which constitute the gravel chosen for our roads, could not be reduced to dust so soon as they now are, by the heavy carriages, if they were firmly bedded in a small quantity of coarse and good calcareous cement, so that the bodies which roll over them should rather compress them, than grind them against each other as they do at present. And as the frequent failures of pavement are manifestly owing to the infirmness of the ground and the looseness of the stones, I have imagined that a solid bed of cementitious work, in the manner of the Romans, and the setting of the paving stones in good mortar,

mortar, would ultimately leſſen rather than enhance the expence. I offer theſe conjectures in the hope, that no body will preſume to decide on the ſubject, who does not know the difference between the common mortar, and the beſt that can be made of lime and ſand; and that ſome public-ſpirited man will make the experiment, where lime is cheap and the expence of pavement or of gravel is conſiderable. If the expence ſhould be found too great for any public works of this kind, the ſame meaſures may nevertheleſs be tried in private areas and walks, in which the neatneſs, duration, and prevention of vegetation, may compenſate for the extraordinary price.

THE END.

www.ingramcontent.com/pod-product-compliance
Lightning Source LLC
Chambersburg PA
CBHW031737230426
43669CB00007B/380